The Found and the Made

The Found and the Made

Science, Reason, and the Reality of Nature

Dan Bruiger

Routledge
Taylor & Francis Group
NEW YORK AND LONDON

First published in paperback 2024

First published 2016 by Transaction Publishers

Published 2017 by Routledge
605 Third Avenue, New York, NY 10158

and by Routledge
4 Park Square, Milton Park, Abingdon, Oxon OX14 4RN

Routledge is an imprint of the Taylor & Francis Group, an informa business

© 2016, 2017, 2024 by Taylor & Francis.

All rights reserved. No part of this book may be reprinted or reproduced or utilised in any form or by any electronic, mechanical, or other means, now known or hereafter invented, including photocopying and recording, or in any information storage or retrieval system, without permission in writing from the publishers.

Trademark notice: Product or corporate names may be trademarks or registered trademarks, and are used only for identification and explanation without intent to infringe.

Publisher's Note
The publisher has gone to great lengths to ensure the quality of this reprint but points out that some imperfections in the original copies may be apparent.

Library of Congress Catalog Number: 2015019094

Library of Congress Cataloging-in-Publication Data

Bruiger, Dan, 1945-
　The found and the made : science, reason, and the reality of nature / Dan Bruiger.
　　pages cm
　Includes index.
　ISBN 978-1-4128-6250-9
　1. Science--Mathematical models--Philosophy. 2. Mathematical models--Philosophy. 3. Cosmology--Philosophy. 4. Self-consciousness (Awareness) I. Title.
　Q175.32.M38B78 2015
　304.2--dc23
　　　　　　　　　　　　　　　　　　　　　　　　　　　2015019094

ISBN: 978-1-4128-6250-9 (hbk)
ISBN: 978-1-03-292187-7 (pbk)
ISBN: 978-1-315-13207-5 (ebk)

DOI: 10.4324/9781315132075

Contents

Acknowledgements	vii
Preamble: The Barcode of Nature	ix

Part One: The World as Found

1	What Is Found?	3
2	What Is Nature?	19
3	What Is Science?	41
4	Law, Chance, and Necessity	59
5	Mathematical and Physical Reality	79

Part Two: The World Remade

6	Consciousness and Its Consequences	101
7	What It Is Like to Be an Intentional System	109
8	The Rebellion against Nature	117
9	The Ideal of Perfect Knowledge	129
10	The Scientific World	139

Part Three: Maker's Knowledge

11	The Book of Nature	155
12	The Religious Origins of Science	165

13	Deductionism, or the Proof Shall Make You Free	183
14	Ideality	199
15	Is Nature Real?	217

Part Four: Beyond the Mechanist Faith

16	Is Reality Exhaustible in Thought?	229
17	Mechanism and Organism	239
18	Theories of Something	259
19	The Next Revolution in Physics?	273
20	The Stance of Unknowing	293

| Index | | 305 |

Acknowledgements

I wish to express my thanks to Caroline Russomanno and the other editors of Transaction Publishers for their painstaking work, with special gratitude to editorial consultant Harold Schneiderman and publisher Mary E. Curtis for taking a chance.

"I was raw; then I was cooked; then I was ash."
—Jalāl ad-Dīn ar-Rūmī

Preamble: The Barcode of Nature

> *"Surprise is inherent in the structure of the world."*
> —Lee Smolin

> *"Nature resists imitation through a model."*
> —Erwin Schrödinger

The next generation of physicists may well produce a set of equations to fit neatly on a T-shirt, claiming they represent a complete and final "theory of everything." This book will show you why no such theory will ever definitively capture the depth of physical reality, let alone tell us how to relate wisely to the natural world. The reason is simple: the very hallmark of natural reality is that it *cannot* be neatly reduced to concepts or equations. There is an unbridgeable chasm between reality and thought, between the found and the made, between natural phenomena and their mathematical models. This is the same categorical gulf that has long plagued the relationship between the physical and the mental.

If nature is *real* it cannot be completely mapped in thought, words, equations, or other idealizations. It cannot be assimilated to human purposes without skewing our relationship to it. While scientific models and other human creations are simple and well defined in principle, the natural world may be indefinitely complex and inevitably ambiguous. If so, there can be no complete expression for the world simpler than the world itself. There is no barcode for nature, no formula for the world, and no realistic hope to stand apart, even conceptually, from the universe that produced us and continues to enfold us. Yet this is not an essentially negative conclusion. For the best hope of humanity, and even for science, is to outgrow a deep-seated denial of nature's fundamental autonomy.

Modern science is haunted by the same ambiguity that has always permeated human consciousness: the ambiguity between internal images and real external things. Despite its self-assured pronouncements, science is plagued by similar questions of the proper relationship between the knowing subject and the objects of knowledge. This book reveals how and why it substitutes its own constructs for natural realities, shaping our relationship to the natural world while limiting the scientific enterprise itself. Like the civilization to which it belongs, science attempts to reconstruct nature with a humanized face. It treats the natural world as a matter of its own definitions, as comprised of simple idealized mathematical models, or as reducible to the parts of a conceptual machine.

The mechanist vision produced the Industrial Revolution, the miracle of technology, and the modern globalist economy, but at the cost of denying immanent reality to nature itself.[1] Far from passé, it has grown more subtly entrenched in the computer age. Yet it fails when considering the natural world as a complex whole, a universe capable of producing life. Like the civilization behind it, science has reached an impasse by following a one-sided approach to the natural world as something to probe, to use, to corral in thought, to reconstruct through technology. Slowly we are realizing that we must embrace a different attitude if we are to survive: we must grasp that nature contains and defines *us*, not the other way around. This book is about this impasse and the parallel attitude that science too must embrace if it is to continue to serve humanity.

The challenge posed by modern cosmology is not only that the universe as a whole presents a unique object of study. What applies to the whole should apply as well to the part. Yet methods of study often ignore the connections of part to whole and usually ignore the role of the scientist in identifying and interpreting those parts and connections. Physics in particular is still colored by a mechanistic vision of matter, dominated by outmoded concepts such as determinism, reversibility, equilibrium, and the isolated system. To embrace the whole, through a more organic metaphor, requires that physics outgrow its one-way relationship to physical reality, based on a straightforward, un-reflexive description of objects. It must become a "second-order" science, capable of embracing its own subjectivity by reflecting upon itself. It must restore a reciprocal relationship with the natural world—and also with the concerns of ordinary people who look to it not only for the latest technology but also for guidance as a secular myth. We have more technical knowledge than at any time in history, but hardly more understanding, let alone wisdom.

Every venture has its ground rules. Permit me here to introduce this fundamental premise: *all cognition is a function of both subject and object*. Everything that one can experience, think, feel or even imagine is shaped *jointly* by the real external world and by one's own being. While this is rather common sense, not everyone will agree. There is a venerable tradition of philosophical idealism, for example, that holds the world revealed by the senses to be illusory, claiming that what is "real" is the mind, the spirit, or some transcendent nonmaterial realm. Many religious beliefs, and those of some scientists and mathematicians, are based on this assumption or something like it. On the other hand, hard-core materialists may deny significance to the subjective aspect of experience and thought, seeking to reduce mind to logic, computation, chemistry and physics. While physicists might be excused for taking this line, even psychology was dominated for much of the last century by a behaviorism that ignored subjective experience altogether. Yet the coevolution of subject and object in human cognition must lead us to include science among the interactions of a biological organism with its environment.

Attempts to resolve the conflicts and swings between realist and idealist poles of thought have nearly defined much of modern philosophy. What I propose is no middle ground but an understanding of a dialectical relationship, which grows out of the ambiguity in consciousness itself. If the fundamental premise set forth above is true, one can see why there is so much trouble surrounding consciousness and what has come to be known as the *mind-body problem,* the *explanatory gap*, or the *"hard problem" of consciousness*.[2] If everything that can occupy consciousness (and even nonconscious mental processing) involves a coproduction of self and world, the situation is very much like a single equation with two variables. Basic algebra tells us that to solve for either of these "variables" requires a second equation—or else to arbitrarily hold the other variable in check. In life, however, there is nothing constant and no second equation. Science generally copes with this situation by artificially bracketing the subjective factor.

Given the basic premise, the metaphor is complicated by the fact that one cannot assume even the concepts and entities of mathematics to be purely mind independent. These too must have been shaped by the needs of a living organism interacting with a real world. Mathematicians may disagree, especially if they share the opinion (early articulated by Plato) that mathematics occupies a transcendent realm with a status independent both of physical reality and the mathematician. This opinion exemplifies a persistent thread of idealism in scientific thought as well, to which new lease has been granted by digital computation. For many contemporary physicists, the underlying essence of physical reality is "mathematical structure," "computation," or "information." A modern theme recurring in physics, as in science fiction and the wider culture, is that we live in a digital, computational, mathematical, simulated, or virtual universe. This theme is steeped in ancient idealism as well as in modern technology. It goes against the ostensibly materialist grain of science. Yet it remains fair to ask: What motivates the apparent need to reduce experience to one or the other factor in what is manifestly a joint affair? Why has science been identified so closely with reductionism, determinism, and mathematics? Why do we even bother to try to contain nature within specific definitions? And how might it serve the interests of science—as well as the world at large—to now pursue a broader approach? These are some of the larger questions this work poses and attempts to answer.

Most people will agree that a primary function of the nervous system is to monitor and engage an apparently real external world that bears serious consequence for creatures living within it. A little acknowledged corollary is that "realness" is how we subjectively experience this relationship of dependency on that world. That is, realness is not only an attribute of the external world, it is also a *quality* imbuing our *experience* of that world. The experience itself of realness has an evolutionary meaning, which refers to that dependent relationship whereby external things portend vital consequences for embodied organisms.

A secondary function characterizes *self-consciousness* in particular: to question the validity of experience, or of conclusions drawn from it, by putting it in the context of being a subjective mental product. This is one role that subjectivity plays, which gives rise to the tail chasing of mind-body dualism. Another role is to provide an inner creative workspace and home base for the self that is sheltered from the contingencies of the external world. The inner subjective realm is already a world apart from nature. It is the ground of all culture and creativity, including the enclaves of science.

These complementary functions have their scientific counterparts. Positivism and operationalism, for example, call into question realist assumptions and conclusions. For science too is a coproduct of self and world. But it is a particular interaction in which the subject factor is "controlled" in order to focus on what the natural world "objectively" is. The subject factor is held constant by standardizing the observer and the repeatable experimental procedure or observation. One is thus better able to rehearse manipulation of the external world by manipulating models of it in thought. Yet scientific method so defined ignores the modeling process itself, and many assumptions behind it, and pretends that the observer does not exist as part of the world observed. Despite its successes, there are limitations to this one-sided approach, which deals with artificially defined parts of isolated systems, to which human agency remains external and unconsidered. As a consequence, agency within nature itself remains ignored, in favor of a mechanistic vision.

The scientific penchant for analysis in terms of mathematical models reflects the cultural influence of the machine and the pragmatic effectiveness of mathematics applied to artificially defined systems. In modern times, these influences merge in the computer, the universal machine, serving both as a powerful new tool and as a new metaphor for reality itself. The computer is psychologically significant because it translates into technology an age-old dream to emulate divine creativity by directly configuring reality. With godlike power one can, at will, command the universal machine to change its very being! Yet it is mistaken to believe we ever will have this power over *nature* in any but limited ways.

The really basic questions tend to involve perennial subterranean veins of thought. While technology shapes our image of nature in every age, the problems of the ultimate whole and of ultimate parts, for example, are archaic. They depend on how one divides up the world, even as reflected in the generalities of logic. While science regularly deals with isolated systems, for example, there are no such things in nature. Indeed, "systems" can only be identified at all in accord with human intentions that only obliquely reflect natural realities. How nature carves itself is quite another question. The current approach, still based on mechanism, treats nature in terms that are clearly matters of human purpose and definition. Yet it is possible to ask how, and in what sense, nature might actively define *itself*.

Because of a long-standing prejudice against active powers within nature, such a question is generally considered inappropriate for scientific study. Consistent with the mechanist metaphor, physics has traditionally dealt with inorganic matter as passively deterministic, lacking inherent powers of self-organization. This bias harks back to the religious origins of science, to early Greek thought, perhaps even to the genetically conditioned psychology of the male psyche. It may be an adaptive trait, insofar as it serves technical mastery. The fact that this approach *works* to produce the technological miracle may seem a strong argument not to meddle with it. However, it does *not* work across the board but applies only to exceptional, idealized situations. It reaches limits in cosmology and in questions involving self-organization generally. It may also have reached its limit as a strategy of the species to the extent it creates a dangerously distorted relationship to nature.

The closest physics has come to acknowledging this prejudice has been to recognize that most natural processes cannot be treated with linear equations and that processes of self-organization within inorganic matter are active at several scales. But science has reached a stage where it can and must give such questions broader scope. The very fact that modern cosmology can ponder the origin and fate of the universe as a whole in scientific terms means that it has reached limits of imagination first recognized by the Greeks. Questions of what there was before the beginning or what lies beyond the edge of the universe continue to defy common sense and possibly to defy falsifiable scientific hypothesis as well. A different approach to natural agency may help in the search for intelligible answers.

In order to carry on, scientific strategy often steers around such impasses, eschewing common sense in favor of logical rigor and formal treatment. Indeed, the recognition of mathematical possibility beyond physical actuality has often led to the discovery of new phenomena. It is reasonable for physicists to believe in the guidance of mathematics. For reasons we will explore, logic and math have been trusted guides to the "object" factor of cognition. Their Achilles heel is that they are nonetheless created by human beings, for subjective purposes. A more balanced and therefore more adequate picture of reality must include the "subject" factor.

Mathematics has facilitated enormous technological success, fulfilling the pragmatic mandate to predict, control, and exploit natural processes for human benefit—thereby confirming the validity of the strategy. However, scientific concepts then tend to become so abstract and foreign to common sense that science fails to fulfill its other mandate, which is to provide society with a comprehensible story about reality—a job once falling to religion.

Scientific interest in nature grew during the European Reformation, as Christian metaphysics mingled with the ancient Greek heritage. Tensions between these traditions had long preoccupied medieval thinkers, who strove to reconcile reason with faith. The conflict between divine freedom and Greek fatalism was resolved by the early scientists in terms of fixed laws of nature, yet with details (now called initial conditions) chosen by divine whim (now called chance). This arrangement translates in modern terms as a *deterministic system* (expressed, for example, as an equation or a computer model) whose inputs are now chosen by the theorist or programmer instead of by God. As descriptions of reality, however, such systems are as mythical as the biblical account of creation. The causal efficacy of equations that might seem to reign through law in nature exists solely on the theorist's blackboard. Scientific models are idealizations serving specific purposes, and can only correspond approximately with reality. "Laws" of nature are essentially descriptive and statistical. They have no "governing" power. Rather, nature simply governs itself in ways that we may or may not recognize as natural laws.

The dream of a complete or final theory assumes that models, if carefully enough made, can correspond exhaustively to that which is found. If this were literally so, however, there would be no difference between reality and theory. The found would *be* the made; nature would be mere invention. Hence, in its extreme form, what I shall call *deductionism* becomes more than a strategy to map the world. It is the belief that the natural world literally *consists* of the conceptual artifacts of science and mathematics. The map *is* the territory. This is sheer delusion, unless nature is wholly artificial to begin with. Yet, of course, that was precisely the original assumption of the devout creationists who were the first scientists!

Reducing nature to mathematical systems plays a useful role and has its place. Yet in the larger picture it may resemble the strategy of the fabled Nasrudin, who searched for his lost key at night under a street lamp, not because he had lost it there but simply because the visibility was better. The key to future science may not lie in the circle of light cast by present science. Illumination comes *from* somewhere; but the location, direction, and color of light—so to speak—are not included in present scientific narrative. The pretension to objectivity feigns a view from nowhere, a uniform illumination pervading all space and time—literally, the vision of the Enlightenment. To advance further, science must reconsider where its illumination comes *from*: its motivations, goals, and basic assumptions. Until present, nature has been considered the passive object of a one-way relationship, a thing to manipulate, confined to human definitions. Science gives lip service to the autonomy of nature as the arbiter of theory through experiment. It may be time to reconsider the possibility that in some way nature is, like us, self-defining and self-creating, at once predictable and unpredictable in the ways that autonomous agents characteristically are. This requires a different attitude than the traditional "objective" stance.

Science gains its technological upper hand by dwelling on the isolated object or system, ignoring the input of both the subject and the rest of the universe. It embraces (and endorses) a third-person perspective that denies cognitive participation and deflects accountability for its preoccupation with single causes and simple systems. As a result, classical physics espoused a view of nature as effectively dead in spite of the obvious fact that nature must be very much alive—at least on this planet—in order to support our own living presence. That our planet bears life does not, of course, imply that the stars, galaxies, and depths of space are literally alive. It *does* mean that the universe must be a certain way to host living observers. Yet the converse is equally important: the embodiment of observers, as physical creatures with an evolutionary history, means that the universe must appear to them in certain ways.

We do not yet fully understand the difference between living and nonliving systems. Recognized self-organizing processes are extensions of known mechanisms. This affords the reductionist promise that we should be able to understand complex, circular, non-linear processes in terms of familiar deterministic models. However, the search for the simple linear relationships behind such models developed for historical more than logical reasons: that's what could be done with the intellectual resources available. We now possess vastly greater resources—the digital computer, in particular—that enable the study of complexity. Yet we are only beginning to entertain a mentality to match.

Such a mentality would tentatively attribute to matter not only the limited forms of self-organization we currently understand, but also the autonomy and power of self-definition that characterize living organisms. The fact that nature cannot be assumed to correspond to *our* definitions does not, of course, imply a *self*-defining cosmos. Yet such an assumption could be as fruitful as its contrary has proven. In any case, matter can no longer be regarded as passively subject to laws imposed on it from without. Laws must be implicit within the universe itself.

No doubt, scientists will continue to struggle with grand questions, including the question of how the universe as a whole could have brought itself into being. However, the shift in attitude that I believe is required to answer such questions is not about a new theory or ontology, which would only continue the outmoded unilateral focus. It is less about what exists than about how to relate to it. It involves exploring how the "subject variable" might be more productively embraced within science. On the one hand, that would mean for science to become, as it were, more self-aware. On the other hand, it would mean that the autonomy of nature would be taken more seriously. While some might argue that science has obviously taken the reality of nature seriously all along, I answer that one cannot adequately approach the object by extirpating the subject. Science has only begun to critically examine its own limiting projections upon nature. Until now it has more or less naively profited by their utility for specific purposes.

Contemporary physics aims toward a theory of everything, and a completed theoretical science is a perennial dream. Yet such a dream is misguided if it assumes that physical reality is exhaustible in thought. Such an assumption implies that nature has no genuine autonomy that can resist complete representation, and no inherent capacity to surprise. It means that there must be a bottom to the complexity of matter and an end to inquiry about the fundamentals, if not the details, of physical reality. Yet to assume that the world is something definite, determinate, with a calculable information content, is to assume that it is an *artifact*, from which we can stand apart. This dovetails with the traditional notion that the subject, disembodied, is no object. Like the creator God, the creative subject has no place in the physical world. This is simply mind-body dualism writ large. The scientific "observer," however, is neither a fly on the wall nor an omniscient being but an embodied and interacting part of the world. This fact places constraints on observation, on what can be known, and on the nature of knowledge. The limitative theorems of the early twentieth century were a first pass at this realization. Einstein made the epistemic circumstance of the observer depend on the finite velocity of light; Heisenberg made it depend on the finite grain of light. Using reason, Gödel, Turing, and others mapped limits of reason itself.

It is safe to imagine that there remain further underlying assumptions that have not yet been thoroughly questioned—for example, the very fruitfulness of mathematics for the study of nature, so taken for granted that many now assume that nature simply *is* mathematics. Platonism aside, mathematics is a descriptive tool. The purposes for which it is used, and its syntax as the language of science, influence our concepts of nature and relationship to it. Yet these influences go unrecognized in the gratuitous mathematization of everything. Even more crucial is the belief—paradoxical for "realist" science—that reality can be contained in definitions at all, let alone in the special idealizations to which mathematics applies.

The needed shift in attitude is elusive, beyond present habits. It concerns not only science—our diplomatic embassy to nature—but social, political, and economic attitudes generally. The domination of nature by humankind mirrors the domination of society by its elites and the domination of women by men. Domination is always justified by supposed truths. Whatever is to be taken seriously must bear the imprimatur of the real! Yet it may be time for even science to outgrow this mentality, which venerates the *object* with too little self-examination. As aspiring gods, ought we not to retain some freedom of choice over what attitude we bring to things?

Perhaps the human need for certainty and justification comes at too high a price. In science, this need is felt in the demand for precise mathematical expression, facilitating prediction and control. It may distantly reflect our species' insecure past as victims of large predatory beasts and tiny predatory microbes—and as victims of each other. As modern civilization takes over the globe, in one sense we are on the brink of universal human rights, institutions, and values. As a common forum for the description of nature, science serves this unification. On the other hand, the unity of science does not imply the unity of nature and certainly not the unity of mankind. The disparity between rich and poor grows exponentially despite liberal ideals, and sometimes because of them. It is facilitated by technology and the mathematization of economics. In part, this perverse effect must reflect archaic aspects of the human psyche: the objectifying, exploiting, and compartmentalizing dynamic that fosters moral disregard for the depersonalized other, while reifying as truth whatever manipulations are convenient and effective for its purposes. Nature has long been such a depersonalized other for us, and only recently has become an object of moral concern.

Apart from social consequences, there is a *scientific* price to pay for the oversimplifications involved in mathematical modeling. Many cosmological speculations are based on tenuous threads of reasoning about rather indirect evidence, yet bolstered by mathematics and sensationally presented as fact. Properties of our universe appear startlingly unlikely as the result of a random shuffle. But the idea of a random shuffle of fundamental "parameters" is a false metaphor that assimilates the complexity of the world to artificial constructs. The world is not a deck of cards or roulette wheel any more than it is a machine; its mathematical improbability is rather contrived to begin with.

Conversely, there are social as well as scientific consequences of mistaking a concept for the reality it represents. Chief among these is devaluing the autonomous reality of nature. For nature is no mere butterfly pinned in a collection but an utter mystery transcending our metaphors and definitions. While our rewarding habit, inherited from religion, has been to regard it as a benevolent provision, it may be time to graciously concede to it—as we have even to corporations—the rights of personhood. Whether that is going too far or not far enough, either way we need nature on our side. In the wake of our major impact on the planet, perhaps nature needs us to behave in a more sporting way. Our planet is the only one we know that harbors life, even life that attempts to mirror in consciousness the ineffable reality of the world. Science is the cultural form officially responsible for that image, on which is based our conduct and attitude toward the planet. Given the circumstances, that is a formidable responsibility.

Many fine books by science writers, and popular books by scientists themselves, recount and explain recent advances (and crises) in cosmology, theoretical physics, and other branches of science. However, this is not my purpose. The goal of *this* book is to scrutinize fundamental orientations of science that have their roots in our very life as embodied organisms, reflecting basic needs, aspirations, and limitations of the specifically human creature. I've written it to fill a void between popular science writing and the more specialized areas of philosophy of science and philosophy of mind. It offers a generalist, critical commentary to informed readers. It does not intend a rigorous argument, but to stimulate thought. It is an essay, in the old sense of the term. Extensive notes and amplifying material coordinated with the text can be found on my web site: www.thefoundandthemade.com.

Science is obviously important to humanity, not only as the underpinning of technology but also as a rational basis for modern culture, as a secular creation story, and as a model of international cooperation. My purpose is hardly to disparage science, which remains for many, as for myself, one of our best hopes. Rather it is to understand its weaknesses along with its strengths so that it may improve; to understand how it both shapes and reflects society's attitudes and practices in regard to the natural world. To that end, it is desirable to identify basic assumptions that underlie both the evolving human relationship to nature and its closely related scientific version. Persistent contradictions within modern science, as within modern society, reflect our inner dividedness as conscious beings. And so it is at that level that we must begin.

Notes

1. The expression *immanent reality of nature* is modeled on the theological notion of *divine immanence*: God in nature. The sense, however, is precisely the contrary: the reality of nature is intrinsic, independent of God, and does not rely on divine presence to fill it from within or command it from without. It includes the sense of the less familiar term *aseity* (self-existence). In particular, the immanent reality of nature contrasts with a derived reality conferred by some external agent, such as artifacts possess.
2. *Explanatory gap* is a term coined by Joseph Levine in the early 1980s to name the difficulties of accounting for phenomenal experience in a materialist framework. *The "hard problem" of consciousness* was coined by David Chalmers in the mid-1990s to contrast with the easier problems of behavioral accounts of cognition.

Part One

The World as Found

1

What Is Found?

> *"God forbid that we should give out a dream of our own imagination for a pattern of the world."*
> —Francis Bacon, The Great Instauration

> *"What we observe is not nature itself but nature exposed to our method of questioning."*
> —Werner Heisenberg[1]

Specifying the Found

Great steam-powered engines fired the scientific and popular imaginations of the nineteenth century, providing metaphors by which to understand natural reality. In the twenty-first century, digital technology fills this role. Our remote ancestors, however, lived closer to the natural world as they encountered it, with simpler technology and metaphors. Fire, food preparation, and animal domestication served as first steps in the disengagement of culture from nature. The distinction between the *found* and the *made* parallels the distinction between the raw and the cooked, the wild and the domesticated, the natural and the artificial.

Such a distinction must bear a different meaning for us than it could have for our ancient ancestors. What is "raw" material for us has already been industrially prepared. The "wild" for most of us is confined to parks and zoos. What we "find" about us in our world is largely artificial, defined and transformed by human enterprise. It is hardly the natural world that might have existed before the rise of our species. Even before agrarian civilization, some socially created world or other served as the milieu into which people were born. The natural world provides us with air to breathe, so that we are not obliged to carry about our own supply like astronauts or those small spiders that live under water. Yet in almost every other respect we mature and live within a bubble of language, thought, cultural institutions, and artifacts. The significance of this becomes more difficult to see as local cultures are absorbed into one global envelope. Yet the question remains: How can one speak of a found world as opposed to the worlds that we make? In other words, how can we identify a realm that is not already transformed by human presence and action, already in some sense manufactured, if only in thought and through language and custom? Such questions are complicated by the fact that we are conscious, subjective beings, aware of our own existence and awareness—and of our complicity in the creation of the human world we take for granted.

The challenge begins with the fact that perception is an active intervention in the world. Even if people had not changed an iota of the world physically, nor redefined it scientifically, we would still be caught in a paradox: there is simply no "direct" access to the raw face of the world. The human nervous system has already transformed its presence in cognitive processes of one sort or another. The senses are not open portals on the world. Rather, it is as though one's brain, sealed within the skull, is wired to the outside world with various remote sensors, which must be interpreted. We cannot get "outside" to witness the world as it "really" is. And, we are burdened (and privileged) with a *point of view*—literally the body's perspective in space and its life in time. Neither science nor human consciousness in general can legitimately claim a "view from nowhere"—a perspective that transcends all perspectives. Thus, what is "found" is unavoidably a matter of context and interpretation.

Barring access to "the world in itself," we subjective moderns might turn to what philosophers call the *phenomenal world* as a point of departure: what we find immediately in our inner life as conscious beings. If one does not come upon *reality* immediately, at least one comes upon personal *experience*. However, the notion of a continuum of experience—a stream of consciousness—is already ambiguous.[2] It could refer either to an unreflective state or to contents of attention earmarked as such. One might prefer to think that the found world is what is passively and innocently encountered in raw sensory experience, unadulterated by reflection. But the very idea of the found already suggests searching as an act, and someone who does the searching and the finding. In the quest for a point of departure, one cannot ignore the modern understanding that even "raw" sensation is already the product of elaborate neural processes and therefore no starting point at all. One can no more turn to "sense data" than to the external world itself to serve as that which is found; much less can one turn to scientific entities and concepts. In the end, we know that all such notions are products of our feverish brains. If consciousness seems to us an output of neural activity, the concept of neural activity is part of that output and hardly an origin. The whole thing chases its own tail!

The Equation of Experience

To attempt to unravel this Gordian knot would take a whole library of books and likely end in frustration, for *wherever* one begins, there will be some corresponding dilemma. The circularities involved are endemic to self-consciousness. Since there is no way out of this hall of mirrors, I will simply begin by naming it the *equation of experience*. This epithet conveys the notion that everything that enters consciousness is a function of two "variables": *subject* and *object*. In this context, "experience" includes not only feeling and thought, along with sensory awareness, but also every form of mental or cognitive activity present to consciousness. Specifically, it includes scientific investigation and knowledge. The "equation" is also a general statement about the epistemic situation of embodied beings, including scientists and their possible computational surrogates. It simply proposes that all cognitive activity is inevitably *a function of both subject and object together*. This is so whether cognition in this broader sense is understood as personal experience, as a form of behavior, or as scientific speculation.

Alternatively, one could take *nature* to be the found world, on the basis that it was here long before us. Yet, that is a modern idea less than a couple of centuries old, with which even some contemporary people do not agree. After all, one cannot *prove* that the world did not spring into being a moment ago, full-blown in its present state—or perhaps in 4004 BC! Even if that were the case, the equation of experience still tells us that our very ideas concerning when it came into existence must be formed *jointly* by whatever the universe objectively is and by our own thought processes. One cannot eliminate one's own input into the equation; one can only ignore it—or disregard it by convention, to better study the other "variable." That is the path chosen by science. Yet the balance between how the variables are weighted is always contentious so that there is rarely universal agreement on what is found, even in science. The equation reminds us that both subject and object are always co-responsible for *whatever* one takes as a starting point.

Hence, there is no hotline to reality, bypassing the conceptual input and epistemic circumstance of the observer. All knowledge, even scientific, is mediated, relational, interpretive, and interactive. On the other hand, there can be no experience or concept that fails to refer, however obliquely, to the world given through the senses: all knowledge and experience is grounded in the body and in physical reality. There are no pure fantasies or fictions, and no pure abstractions—nothing invented without a basis in the external world.

The two variables in the equation always act together, and there is no second equation, as it were, to help "solve" for a single variable at a time. One cannot expect to know the world in a purely objective way, for knowing already is an act of the subject. On the other hand, mind (even artificial mind) that is not a physically embodied product of an evolutionary process is not a real possibility. The equation of experience precludes mind disembodied, or isolated from the external world, just as it precludes meaningful talk of a universe without observers. We are free to *imagine* either, but neither characterizes our actual existence.

The very notion of objectivity suggests that the world can be known independently of the knower and the process of knowing. *Natural realism* thus presumes that the world has a reality apart from how it is known. Thus science initially proceeded by conventionally excluding the role of the observer in order to affirm the reality of the world. One difference between classical physics and quantum physics seems to hinge on this very issue: Classical properties are thought to inhere in objects themselves, while quantum properties implicate the role of the observer as well. Yet even in the classical realm it is only scale that permits the role of the observer to be disregarded.

The large size of ordinary objects in relation to the medium of investigation (e.g., photons) permits effects of the latter to be ignored, so that the object can be considered in its own right. This circumstance of scale is an incidental fact of the world we live in, to which we have adapted with an appropriate stance we call realism. While this stance works, in context and for its purposes, we cannot count on it to be valid in every circumstance. The reality of nature as a whole, therefore, cannot be embraced from the restricted point of view of classical "realism," whose concepts are actually severe idealizations only marginally in tune with the wholeness of nature. While such idealizations purport to represent the objective reality of the systems studied, they draw liberally upon the power of the subject to redefine nature at will.

It might seem that quantum weirdness does not support the reality of nature in the way that classical physics appears to. If anything—we shall see—the very opposite is true. For, the idealizations of classical physics, however rational, are actually fictions, while the irrationality of the quantum realm is actually the hallmark of nature's reality. Furthermore, in *both* macroscopic and microscopic realms, *all* phenomena are essentially relational and statistical, a result of repeated measurements, in which the observer is always implicated.

The question of whether quantum mechanics describes the world itself or rather our *knowledge* of the world seems moot. We shall see that it is no less so for classical mechanics. Perhaps, for psychological reasons, we must conceive the electron as an object, even though it functions as shorthand for a complex of measurable quantities, such as mass and charge. In any case, the electron's reality—whatever that is—is a different affair than its formal representation in theory; one is found, the other made.

Though both are formalisms, quantum physics differs from classical physics in being driven by observational results that seem irreducibly statistical; the data themselves prevent an interpretation in terms of ordinary objects. Quantum physics is thus profoundly empirical, if not realist. It originated in the first place because of the failures of classical theory to match empirical evidence in the realm of high energies and small sizes. Bohr insisted on the inseparability of subject and object in that context, against Einstein's "realist" insistence that the quantum theory could not be considered complete until it allowed the sort of prediction of individual events permitted in classical theory. Ironically, this expectation is precisely what is ruled out by the reality of nature. It is an expectation based on the power reason holds over its own constructs—which is not realism at all, but an idealism that assumes an impossible omniscience.

Physical theory makes no fundamental distinction between elementary particles that occur naturally and those that occur only as by-products of high-energy experiments. In that context, the made is not distinguished from the found. The range of entities incorporated in the standard model of particle physics[3] includes whatever is allowed by theory, regardless of whether such things naturally occur. To many, this suggests other possible worlds, with varied parameters permitting various theoretical entities. Unification of such possibilities within a common framework thus amounts to the unification of the natural and the artificial, in a theory that deliberately blurs the distinction, so that all things are subsumed as functions of human definition.

Descartes and Husserl

As the Greeks had acknowledged, one could be misled by the testimony of the senses, which depends on the limited and corruptible body. What could *not* be doubted, said Descartes, was the sheer *fact* of experience.

He applied the latest anatomical knowledge to ancient skepticism regarding the vulnerability of the senses. Suppose that an "evil genius" had limitless powers to alter or manipulate the nervous system in such a way as to completely fake the testimony of the senses, even giving the sham appearance of a real external world.[4] Still, he argued, while one might doubt the truth or interpretation of this testimony, one could not doubt its occurrence.

While Descartes sought an unassailable high ground in the fact of "phenomenal" experience, it was Edmund Husserl who much later coined the term *phenomenology* and aimed to chart its domain. Even to view experience *as* experience, he pointed out, requires a special attitude, which he called *bracketing*.[5] This was supposed to be a way to regard the content of experience without layering any interpretation upon it. Even better, it was a way to observe oneself in the act of imposing such interpretation, to study mental processes from the inside, so to speak.

The default state of consciousness is simply awareness of the *world*, which Husserl calls the *natural standpoint*. This is also the usual standpoint of science. As Kant had recognized, however, perception invariably involves cognitive judgments. While acts of the organism, these judgments are normally projected as aspects of the external world. This means that there is a projective faculty of mind normally in operation, which is responsible for our perception of externality—the fact that we normally experience the world as "out there," not as unfolding on the body's sensory surfaces. It leads also to perception of the world as *real* and to the tendency to objectify or externalize all sorts of ambiguous and possibly subjective traits as concrete aspects of the external world. Accordingly, one's fundamental experience of the world is of real things separated and localized in space outside the body and of real events distinct in time.

As human creatures, we are natural realists; we normally perceive the *world*, not some inner domain. For the most part, this arrangement works well. Hence, experience must be deemed to mirror the structure of the external world, at least on the human scale: "objects" really exist, as they are given in perception, and can impinge on us in ways that affect our well-being. Though reflecting reality in such a way as to favor survival, the "realizing faculty" is an organizing principle imposed on sensory input, a necessary if not sufficient condition for the life of the organism. It may not be equally helpful in all circumstances or at every scale. Indeed, it can lead individuals and societies into deep trouble when their own psychic attributes are projected as features of reality. This ability is thus a double-edged sword. Because it bears potential dangers for a highly social creature, we might expect some corrective function to have evolved as well. This, I believe, is one important role of reflexive consciousness.[6] As we shall see, that is not its only role, for the inner world becomes the laboratory for every sort of invention and mental creativity. It is the realm in which language, abstraction, and artifice are possible.

Realism and Positivism

The natural standpoint is the natural basis for scientific realism. It constitutes a first order of description, reflecting the normal focus on the world "out there." Husserl points out that this focus can be relaxed in the psychological act of bracketing, a function of consciousness that establishes a major freedom of the self. To resist the compelling persuasiveness that often imbues experience is to defy the natural standpoint. While the external world normally and justly claims attention, against the possible tyranny of this claim subjectivity asserts the need to question the veracity of experience—that is, its relation to the external world. This is the basis of a second order of description, which must include the relationship of the world to the subject.

Bracketing some aspect of experience (or experience on the whole) involves a suspension of belief. The scientific version of this is often called *positivism* (though ironically it negates assertions of the realizing faculty). Similarly, *operationalism* considers science to yield results of *observation* and *measurement*, rather than direct knowledge of real entities. Properly, such "skeptical" positions reflect the need to question the role and status of specific scientific concepts in particular contexts and to focus on the procedures involved in creating such knowledge. The observed properties of things, which a positivist might emphasize as the reliable basis of knowledge, are not defined only by relationships among external things but also by their relationships with the observer, as established through such operations as measurement. This does not imply that things *have* no intrinsic reality or real properties, only that knowledge of them involves actions of the observer.

Positivism seeks grounding in the immediacy of perception. In polar contrast, the deductive thread of science seeks assurance in the logical consistency and creativity of ideas. These two great themata[7] were in contest in scientific thought at the turn of the twentieth century; in them we see the interplay of the two "variables," subject and object.

In addition, we see the human inclination to prefer one to the other, rather than to admit their codependence. Each position bears a partial truth: they are not inherently contradictory, and science must embrace both to be complete. Because that is humanly challenging, they seem fated to alternate as dominant themes in a dialectic cycle.

Primary and Secondary Qualities

The notion of sensory *quality* has both a subjective and objective aspect. It refers to phenomenology, on the one hand, and to material properties, on the other. It belongs ambiguously both to the self and the world, testifying to their interaction and joint authority in perception.

The early scientists believed that the size, shape, and motion of objects were "primary," or objective, qualities. In contrast, taste and odor, for example, were "secondary," or subjective, qualities—mental responses to sensory stimulation by the object's material properties. Though primary qualities may be considered properties of external things, they are nevertheless relational. Distance, for example, is in relation to some reference point. Physical properties have been treated variously according to whether they are thought to inhere more in the thing or in the relation. The properties of classical objects seem to inhere within them, whereas properties of a quantum system cannot even be defined without specifying the experimental setup.[8]

As Husserl notes, the distinction between primary and secondary qualities is rather contrived. To subtract so-called secondary qualities from a phenomenal object does not yield pure extension or geometry, for example, any more than erasing colors from a photo yields a coloring book outline. What remains instead is not even the physical space we know in perception, but—if anything—a complete abstraction. Like the ancient philosophers, scientists are at liberty to invent abstractions to propose as the "true" fundaments of the physical world. No one can claim, however, that these are natively found.

Some qualities in particular appear to demonstrate intentionality. In contrast to "neutral" qualities like color and auditory tone—which are functions of the distance senses—pain, fear, anger, and hunger, for example, are clearly associated with motivated behaviors. They have teleological (and hence the evolutionary) meanings implied in the *actions* associated with them. The act of bracketing can dissociate the experienced quality from the associated action that reveals its natural intent. But qualities like color and auditory tone are not associated with an impulse; they bear no obvious intention, and do not seem to imply an action or judgment. What is the behavior implied by color or auditory tone that would shed light on the "meaning" of these as experiences?

One could speculate that seemingly neutral qualities might once have been associated with an impulse. There might in our evolutionary past have existed referents of an emotionally charged sort for specific colors and sounds: blood, for example, for the color red; or dangers associated with specific auditory cues. Yet the most *general* meaning associated with the distance senses refers to the *externality* of the world. Events in the visual and auditory fields are normally taken to indicate the presence of real external things. The affective content of such externality lies in the need to take the environment seriously—as holding the power of life and death over the organism. This is the *meaning* of realness as an experience and as a cognitive judgment. It reflects the intentionality manifested by the distance senses, whose mandate is to identify and give advance warning of the nature and location of potentially significant "objects" impending in the environment. These could be threat or prey, to be navigated as obstacles or used to advantage. As well as an attribute of the world, then, realness is also a specific *quality* with which (some) objects of experience are imbued—like the redness of blood or the hurtfulness of pain.[9] It refers to the fact that matter necessarily *matters* to the embodied creature. Through the panorama of experience, at least the human creature narrates to itself an ongoing account of its highly interested relation to the world.

While this may seem patently obvious, it is laden with consequence for the organism, which suggests that realness itself must bear a kind of psychological charge. This is where the quality of intentionality most naturally has its sense of referring *to* something.[10] Even if a sensation does not contain within it a cognitive judgment implying a specific action, it may nevertheless contain the cognitive judgments of externality and realness that generally attend perception of the world. The sense of realness is charged with the organism's implicit knowledge of the powers of real things to affect it; yet it generally goes unrecognized as a cognitive *action* by a realist outlook that passively views the world as through an open window.[11]

The scientific disenchantment of the world reduced the richness of phenomenal experience to measurable primary qualities (i.e., quantities), while confining the "show" of experience to the interior of the skull. This reduction is one symptom of unresolved dualism, in which the paradoxical relation between image and reality is not properly understood. It demonstrates the circularity built into cognition conscious of itself: the constructed world of physics is recycled as the reality held to underlie both physics and the world constructed in ordinary perception. Most people do not seem to suffer greatly from science's disenchantment of nature, preferring rather to benefit from its innovations. In any case, physics' image of the world is not necessarily *truer* than that of ordinary perception; rather, it serves a parallel function on a different level, and plays by many of the same rules.

Things or Patterns?

Objects in science play a somewhat different role than in ordinary perception, especially in contexts that lie outside the range of the latter. In the microscopic realm, "particles" are small entities that do not behave like ordinary things, and do not even seem to be individually identifiable. On the largest scale, the universe as a whole can scarcely be treated as an object in an ordinary sense. Yet, natural objects are ambiguous even on the human scale. How well defined is a tree? What are the boundaries of a cloud, a mountain? In the face of inherent ambiguity, it is understandable that the mind should impose its own clearer boundaries and definitions.

Science overcomes this natural ambiguity by simply *defining* its entities to begin with. Thus, scientific entities are made, not found; they are not natural things but *products of definition*. As such, they are precise in essence. Their well-defined existence within theory can only correspond approximately to the comparatively fuzzy results yielded through the interactions we call measurement or observation. Definition produces useful schemata to account for observed phenomena. Yet these are often also treated as the literal realities *behind* the phenomena concerned. We tend to think of atoms as small objects, as solid and real like the tables and chairs they compose, despite also believing that they consist of mostly empty space. This reflects the evolutionary success of our perceptual strategies. But if tables and chairs behaved according to quantum rules, we might be less inclined to perceive them as solid or real.

Are there literal objects at *any* scale? Or are "objects" merely shorthand for recurring patterns? Either way, there must be criteria to distinguish patterns that correspond to real structures in the world from a background of other appearances that do not. Invariance plays the role in science that perceptual constancy does in ordinary experience. Objects are consciously inferred in science in a manner parallel to how they are inferred unconsciously in ordinary cognition. And, in both domains, this inference entrains an inescapable circularity. If cognitive processes deduce "objects" from "data"—whether gathered from sensory surfaces or instrumental detectors—one is tempted to ask what causes these data in the first place, if not real things in the environment? How is *input* of data to be conceived, if not as it is conceived in the *output* of the cognitive process?

Such embarrassments arising in the scientific picture of the world have the added complication that objects there may defeat expectations based on ordinary experience. One wants to know, for example, what real things cause micro detection events. Notwithstanding the circularity, in ordinary experience one has at least the illusion that one can often point to the real things that are held to cause our perceptions and conceptions of them. This is hardly so in the quantum realm, where things cannot be pinpointed to a definite location, where measurement may alter what is found, and where it may not even be possible to verify the presence or characteristics of entities between sightings. (How, for example, to spy upon a photon traveling between its point of emission and point of absorption?) One cannot count micro *objects*, but only the macroscopically visible measurement *events* they are presumed to cause in detectors, such as the photoelectric current generated by light or the blackening of grains in a photographic emulsion. These events may not represent identifiable "things" at all, but merely quantities of "energy." This circumstance has led some to conclude that quantum entities do not "really exist" between measurements, or that the question of such existence has no meaning.

While scientific concepts often depend metaphorically on ordinary experience, there is an influence both ways. Our modern everyday notion of "objectness" draws heavily upon industrial artifacts and the scientific concepts behind them. A billiard ball is a far more archetypical object for science than a stone or a pinecone, let alone a clod or a cloud. Industrial objects are well defined and functionally precise. This familiar crispness of made things carries over psychologically, if not logically, to our ideas about natural things. Yet it is a mistake to attribute to natural things the precision by definition of artifacts. Though we may imagine electrons as universally identical ideal objects (like billiard balls), there is no corresponding way to inspect and compare them as individuals.

Counterfactual definiteness is the notion that there must be a fact of the matter concerning something regardless of whether a measurement or observation is made. It serves to clarify the meaning of realism and its measurable consequences. Quantum effects, for example, should be distinguished according to whether they are found or made in the course of the experiment.[12]

In general, measurement produces a number that is meaningless apart from a theory that specifies *what* to measure and what a measurement *is* in a given context. For example, theory might tell us whether to count discrete events (clicks of a Geiger counter) or to inspect an analog scale (a movable pointer). In any case, a *series* of measurements is required. For a measurement or observation to be meaningful scientifically, it must be repeatable, so that measurement is intrinsically statistical. The universe does not generally stand still for the observer, however, so that only relatively stable or coarse-grained features can repeatably be measured. A large-scale pattern or regularity is earmarked as the genuine effect, while fine departures from it (the "error bar") are considered random fluctuations—noise. An effort is made to separate the error due to the apparatus from other sources, but in the quantum realm this distinction is not always feasible.[13] And what is at first considered noise can turn out to have a deeper significance.[14]

Powers of Language and Abstraction

Language expresses the power to invent and the power to dissemble. It allows one to substitute a symbol for a real thing. While the mind-independence of physical reality implies that we are not at liberty to change it arbitrarily, the essence of language resides in this very possibility. The fact that one can make grammatical statements that are not true, or even semantically meaningful, is what gives imagination expressive license, both in word and in deed. The inventiveness of language is closely related to invention in technology.

The human abilities to abstract and to imagine turned the natural order upside down. Reality could then be an invisible, immaterial realm of ideas and symbols, rather than the evident, tangible, material presence of nature. Abstraction allows us to organize experience into categories and to perceive relationships that lead to generalizations with both predictive and creative power. While such generalizations may refer to real phenomena, the fact that they are mapped by symbols, which can be freely manipulated according to formal rules, allows for counterintuitive assertions. It also implies that imagination can rearrange physical materials in new configurations, with their own syntax, so to speak. Language is closely associated with toolmaking, involving overlapping cortical areas in the brain. The same power of abstraction permits natural principles to be applied to create artificial systems: machines both physical and conceptual. The complex interplay of syntax and semantics in natural languages was itself distilled in the concept of the *formal deductive system*—an abstract version of the machine.

It fell to Plato to propose a truly abstract (read: ideal or spiritual) basis for material reality.[15] It is probably no coincidence that such abstraction first occurred to the West in ancient Greece. Greek language and thought influenced medieval Christianity, first through the translation of scriptures into Greek, later through the rediscovery of Aristotle in Arabic translations from the Greek. A characteristic of Greek, as of English and Indo-European languages generally, is that the verb *to be* serves multiple functions (which are differentiated in other language families), allowing mere predication to imply existence.[16]

Words are thus not just afterthoughts to name concepts, but are instrumental in forming them.[17] Language gives an independent autonomy to the names by which things are known; indeed, it helps to confer "thingness" upon sensory configurations. It reflects and recapitulates—but also shapes—the cognitive schemata through which we experience the world. Words and symbols are not mere tokens or signs incidentally associated with external things; they do not merely name but also create.

Language and thought take a dramatic further step when words become identified with precise meanings through formal definition. Through the act of *defining*, their meaning is deliberately confined to something specified in given words, as opposed to the diverse things they have been commonly *associated with* through general usage and experience. They then no longer have an ambiguous relationship to the external world, with several meanings in various contexts. Through definition, they become *exactly and only what they are determined to be* by explicit consensual agreement. They are then no longer references to found things, but become instead references to things within a constructed world, which exist by the fiat of definition. We shall see that this is not only the basis of scientific modeling but the very foundation of consciousness.

These properties of language facilitate the general tendency of thought to *reify*: to conceive processes as things, making verbs into nouns, so to speak. They give closure and a self-contained quality to the domain of thought, since mental acts become new objects for further mental acts, independent of the real world. In ordinary language use, one does not expect meaning to arise from syntax alone but from semantic reference. In science the matter is not so clear. For, reducing phenomena to mathematical expression essentially reduces them to symbols connected by syntax. The elements of such an abstract "world" are the symbols themselves, not their referents; its laws are first of all rules of grammar, and only indirectly empirical patterns. This is the basis of deductive systems.[18]

The conceptual world of physics is self-contained in this way to the extent that its elements are idealized products of definition, corresponding directly to nothing in sensory experience, and only indirectly, symbolically, and approximately to anything in the real world. This autonomy, we shall see, exploits an enduring preference for deductive systems, a general tendency to prefer the certainties offered by man-made systems to the ambiguity and contingency of the natural world. It behooves philosophers and anthropologists to study how variation in the structure of languages gives rise to diverse ways of thinking, and to pay attention to the historical influence of particular linguistic structures on Western experience in general; it behooves scientists to pay attention to this influence on the development of scientific thought.[19]

Like ordinary life, science cannot proceed without the use of metaphor. Its business, after all, is to extend the reach of thought and sensation beyond the familiar realm to which language is naturally adapted. Our vocabulary and terms of reference are framed in terms of the macroscopic world, which provides a basis for our categories of thought. Therefore, it is normal to frame concepts outside that domain in terms of things familiar within it, which is the function of metaphor. Apart from mathematical description, or a language of pure abstractions, there is no other way to speak about phenomena that are beyond human perception because they are invisible, too small, too large, too far away, or too complicated for detailed comprehension. Thus, physicists continue to speak of "waves," even when they are not considered waves *in* some medium, and of "particles," even though they are not considered solid things in the ordinary sense.[20] Such extension is always risky, since it constrains us to think in the limited terms of familiar images. Combined with the properties of language discussed above, this can lead to taking the metaphor too literally. In particular, the metaphor can be mistaken for what it represents, the map confused with the territory.

Notes

1. Heisenberg, Werner. 1989. *Physics and Philosophy: The Revolution in Modern Science*. New York: Penguin, 1989.
2. Some have questioned whether there *is* such a continuum at all. For instance, Blackmore, Susan. "The Grand Illusion: Why Consciousness Exists Only When You Look for It." 2002. *New Scientist* 174:26–29.
3. The current very successful theory of fundamental physics, uniting the strong, weak, and electromagnetic forces and classifying the known particles.
4. The classic "brain in a vat" thought experiment is a variant of this, as is the more contemporary version in the film *The Matrix*.
5. Or *epoché* = suspending judgment.
6. That is, the awareness of awareness; to be distinguished from simple awareness.
7. The notion of themata (plural of *thema*) was introduced by historian of science Gerald Holton.
8. Gomatam, Ravi V. 1999. "Quantum Theory and the Observation Problem." PhilPapers. http://philpapers.org/rec/GOMQTA, sec.vii.
9. Note that "real" and "fundamental" bear different meanings. Wetness, for example, is considered an emergent property of a more fundamental level of description, but it is no less real for that.
10. Following Brentano, *intension* (distinguished from *intention*) gives the sense of directedness or aboutness of many mental states, such as thoughts, feelings, beliefs, and intentions, reflected in language statements. Formally, an *intension* is a set of defining characteristics, while the *extension* is the set of things satisfying those characteristics.
11. Rainer Mausfeld is one experimental psychologist who does recognize it. See Mausfeld, Rainer. 2013. "The Attribute of Realness and the Internal Organization of Perceptual Reality." In *Handbook of Experimental Phenomenology: Visual Peception of Shape, Space, and Appearance*, edited by Liliana Albertazzi. Chichester, UK: Wiley, 98–118.
12. Aerts, Diederick. 1998. "The Entity and Modern Physics." In *Interpreting Bodies: Classical and Quantum Objects in Modern Physics*, edited by Elena Castellani. Princeton, NJ: Princeton University Press, 239–40. This is not always clear, as in delayed-choice quantum measurements, for example. In some cases, the assumption of counterfactual definiteness determines the type of statistics employed, according to what count as real possibilities.
13. Margenau, Henry. 1958. "Philosophical Problems Concerning the Meaning of Measurement in Physics." *Philosophy of Science* 25:23–33.
14. As famously happened with the discovery of the cosmic microwave background radiation.
15. Cassirer, Ernst. 1955. *The Philosophy of Symbolic Forms, Vol 1: Language*. New Haven, CT: Yale, p. 73.
16. Cassirer, Ernst. 1957. *The Philosophy of Symbolic Forms, Vol 3: The Phenomenology of Knowledge*, New Haven, CT: Yale, p. 333. See also note 19.
17. Ibid, p. 331.
18. See http://en.wikipedia.org/wiki/Formal_system. I intend the notion of deductive system less formally, to include any construct formally defined.
19. Dewart, Leslie. 1969. *The Foundations of Belief*. New York: Herder & Herder, pp. 94 and 138–39.
20. Lewontin, Richard. 2000. *The Triple Helix*. Cambridge, MA: Harvard, p. 3.

2

What Is Nature?

"Nature tends to a wild profusion, which our thinking does not wholly confine."
—Nancy Cartwright[1]

"Nature's independence is its meaning; without it, there is nothing but us."
—Bill McKibben[2]

Natural Autonomy

To the ancients, the *nature* of something was its inherent essence: what it is fundamentally, in itself, in the wild. This was the source of its properties, powers, behavior, qualities, and relationships to other things. Nature as a whole consisted of the natures of things collectively. Such a view is quite foreign to modern science, which focuses more on its own theoretical constructs as they relate to controlled experiments and observations. These are *interactions* with nature that are often highly mediated by technology in well-defined artificial situations. The results of such interventions are used to choose among competing theoretical models, which are, in effect, the actual objects of study. While scientific realism accords nature the authority to decide scientific questions, the questions themselves tend to ignore nature's independent reality, especially its active powers of self-organization. The nonliving cosmos is still viewed implicitly as a passive object of attention, curiously resembling the mechanical systems constructed to investigate it. Such systems lack powers of self-organization by definition. As a scientific object, "nature" is now little more than an adjunct of the technology involved in its study, just as it is little more than an economic adjunct of our civilization.

The root meaning of the word *nature*, however, suggests self-generation, a reality that maintains itself independent of outside agents, whether human or divine. While referring initially to birth (as in *nativity*), it has come to mean the whole of physical reality. Primordial, self-generating, and autonomous, nature properly stands in stark contrast to technology and the human world. It is something found as opposed to made. The very meaning of natural reality implies the ability to transcend human thought: in order to be *real*, the physical world must forever be more than what we think it is. Containing *us*, it can only provisionally and partially be contained within scientific models. However, this is not the understanding modern science inherited from antiquity and its medieval religious origins.

Plato and Aristotle had already deflected attention from nature's self-existence by distilling the essence of a thing as its "form." To be sure, Aristotle distinguished natural things from artifacts. The former contained within themselves the power of their own material realization, while this was imposed upon artifacts from without. But Aristotle blurred this distinction by regarding things in general, whether natural or artificial, as form imposed upon substance. He thus subordinated the independence of the natural world to human thought, paving the way for creationism and mind-body dualism, along with scientific determinism and expectations of a final theory.

According to our Christian heritage, the natural world is *not* self-generating, *not* independent of outside agency, and *not* primordial. The notion of nature as created *ex nihilo* emerged with the monotheistic concept of God, signifying the separation of subject and object writ large. Such a nature has only the derived reality of a created artifact. This ignored Aristotle's distinction between the natural and the artificial, while upholding the mental or spiritual power to creatively impose form. God was separate from nature and above it, in the way that the artisan is separate from the artifact, and mind is separate from body. This implied that nature was not the fundamental reality, not self-existent, not divine, and certainly not to be worshipped. On the contrary, it was fair game to study, manipulate, and exploit for human use. Thus estranged from mankind, it was as distinct from the Creator as the body was from the Christian soul. The physical world did not reflect our true being, which is spiritual. Nature was not our true home, which is otherworldly. Rather it was a punishment, the wilderness into which we were cast from the Garden.

The image of nature in every society and era has a political dimension with ecological consequences. As a category of civilized thought, nature represents the wild, the untamed and chaotic, which can pose a threat to social order and political power. The perception of wilderness as a dangerous no-man's-land may have served to rationalize early urban concentrations, as well as to bolster the need for civilized conduct in close quarters.[3] The contemporary image of nature as a remote outback, or a mere appendage to urban life, may serve to reassure modern people concerning the consequences of consumerism and unlimited growth.

Nature as a whole was once thought to be like a living thing, with powers of agency like an organism, not merely the passive inertness to which the philosophy of mechanism has reduced it. In modern terminology, the world was *autopoietic*: self-creating and self-defining, as well as self-maintaining and self-reproducing.[4]

In contrast, the natural philosophers of the Enlightenment held that God imposed order on natural chaos, just as rulers imposed it in the human realm. This identification of human with divine mind conformed to the heritage of patriarchal religion, in which God is a projection of the masculine psyche. Nature was believed comprehensible because the mind of its Creator was reflected in the reason of the early scientists. The rational design of nature could be contemplated and discussed, as from one engineer to another, upstaging nature's manifestly irrational aspects, and bridging the gulf between God and man, intellectually if not morally. Nature's design, and not nature itself, was the important thing: the distilled essence, the blueprint.

From the Greeks on it had been a slippery slide into a mechanistic world. From ancient Greece we inherit *deduction* (of knowledge from first principles) and *reduction* (of wholes to parts). The Judeo-Christian-Islamic tradition contributed the idea of an external creator. Moreover, medieval society was fascinated by machines. Altogether these influences fostered a vision of nature as comprehensible, orderly, and mechanical—an artifact lacking its own intentionality. Nature was *definable* (since it had been literally spoken into existence); hence, it could be definitely known, reduced to a conceptual scheme presumed to resemble what its designer originally had in mind.

The Greeks had discovered the power of deduction, which is the basis of mechanism; viewed abstractly, machine and deductive system are interchangeable concepts. This discovery became the basis of the scientific ideal: to axiomatize every branch of knowledge as a deductive system, based on fundamental definitions. Essentially, to mechanize knowledge. Modern science continues to view inanimate matter as passively obedient to externally imposed laws, much like a computer, rather than imbued with its own powers of self-organization.

The Christian reading of Aristotle had transferred all teleology and creative power in nature to the masculine creator, God.[5] Whereas Aristotle had analyzed nature in terms of four types of cause, which included teleology, the early scientists dropped all but "efficient" cause. Yet somewhere behind passive transmissions of effects there must be an agent to initiate them. On its own, a natural system was inert and could change only through disturbance by something outside it. God was deemed the first cause of a potentially endless chain of effects.

In Aristotle's terms, however, efficient cause did not refer to the natural order of things or the normal state of affairs, but to some unusual or extraneous circumstance or event. Efficient cause was something that provoked nature into an exceptional state. Accordingly, Aristotle held that nature itself is not passive, but normally just carries on according to the natural order of things. On the other hand, to know that order, the observer should take care not to interfere with it. Aristotle cautioned against experimentation as a way of forcing nature into an unnatural state. This could not lead to a genuine understanding of natural processes, he thought, but only to artifacts he considered *monsters of nature*.[6] Rejecting Aristotle, the new science evolved precisely as the study of such artifacts, and experiment was embraced as the means to deliberately produce them.

Aristotle makes his case for teleology in terms of living things, which develop spontaneously through stages of growth. The awkwardness of applying this vision to *inorganic* processes may be one reason why the early scientists rejected immanent final causes in favor of efficient causes and a notion of finality as something imposed externally on nature by divine will.[7] Nature did not *need* self-organizing powers, since it had been *divinely* organized at the outset—an assumption eventually challenged by Darwin. In mind and spiritual essence, man had been supposed to share with God a position outside nature, well situated to tinker with the world like the Lord himself. There was from the start a fine line between probing nature for clues to divine reason and imitating divine creative powers. Both relied on treating matter as passive, inert, and without the powers of self-organization associated with living things.

In many ways, this vision of nature remains current today. Efficient cause remains the underlying basis of determinism, which still requires a first cause in the form of some input from outside a system considered passive. While the role of first cause is no longer attributed to God, either it is ignored, or deferred through further deterministic analysis, or else the theorist steps up to assume the role by writing initial conditions into the equations. The relation of theorist to theoretical model recapitulates that between (divine) creator and creation. The question of the inherent creativity of nature itself remains outside the bounds of this framework and has only been forced upon science by the indeterminism of the quantum realm. The idea that nature is *free*—that is, unbound by causal principles at the deepest level—is "both thrilling and scary," since it "fails to satisfy the demand for sufficient reason—for answers to every question we might ask of nature."[8] One resists the freedom of nature for the same reason one fears the freedom of other human beings, whose behavior cannot be easily predicted or controlled.

The dream of modern physics to achieve a theory with no "independent parameters" is consistent with its vision of nature as lacking autonomous reality; but it is inconsistent with the very notion of reality itself. Only if nature is merely an *idea* could it be completely contained in any theory. String theory is the closest that modern physics has come to this goal. Yet string theory is hardly monolithic in the way that truth is supposed to be; it consists of an indefinite number of possible versions, with no way at present to choose between them.[9] The implicit expectation is that *nature* should ultimately choose the correct version in some form of experimental test, presently beyond reach. Yet how can nature perform this service if it is not ultimately autonomous?

Mechanism and Creationism

The religious heritage of science deeply informed the philosophy of mechanism. Divorcing teleology from nature was a reasonable choice for a budding science that sought to apply reason to the relatively simple phenomena it could embrace. It was also a *theological* choice aimed to preserve divine potency against pagan ideas. A world that could design itself, after all, had no need of a creator to design it, whereas a world designed by a creative god was guaranteed to be commensurable with human intelligence made in the divine image.

Nature would understandably appear mechanical to an age obsessed with the potential of machines. In our own age, the archetype is more abstract: the *universal* machine—the digital computer. Just as the computer generalizes and formalizes the concept of machine, the *effective procedure*, or *algorithm*, generalizes and formalizes systematic treatment—the essence of mechanized operation as the basis of computation. The equations of physics are such algorithms, and one effect of the widespread influence of computers is that nature is now commonly understood in terms of computation.

Mathematics, of course, has been indispensable to science since the time of Galileo, who proclaimed the "book of nature" to be written in the language of mathematics. Yet this utility falls short of claiming that nature *itself* is but a "mathematical structure," or that the physical world is *nothing but* mathematics. While nature can certainly be described mathematically, this is so only to the degree and in the ways that it resembles a machine—that is, a product of definition. Rather circularly, this resemblance comes of a particular way of looking, which focuses on those aspects of natural reality that can be treated mathematically. However, it is not nature that resembles a machine, but a cartoonish vision of nature.

What is Explanation?

How shall one understand the fact that solid objects fall to the ground? Or that it rains more in winter than in summer in some places? These are *patterns*, which one might (like Aristotle) simply accept as the normal way of things. Alternatively, events can demand further explanation when they do not fit into recognized patterns.

An explanation attempts to assimilate something to a larger context, to rectify a cognitive dissonance, restoring a sense of the normal way of things. Only when we are first led to wonder about an event, pattern, or phenomenon do we seek an explanation. For Aristotle there was a natural way that things were supposed to be. Only apparent exceptions to this natural order required explanation. Thus, the phenomenon we now call gravity had been rationalized as the "natural" drift of heavier things toward "the center." For Aristotle, there was simply nothing in gravity to explain. Conveniently, the center of the earth coincided with the center of the cosmos, so that the behavior of falling objects seemed to confirm the geocentric worldview.

Magnetism as a recognizable "force" must have played a role in reinterpreting gravity as a similar phenomenon—an efficient cause that invited an explanation in terms of further efficient causes. Galileo had the counterintuitive realization that objects continue at rest or in uniform motion until acted upon by some external disturbance. The natural state shifted from motion toward the center to what we now call inertial motion. An explanation of gravity then required the larger context of *forces* to effect changes in inertial *state*. Newton grasped that the same (invisible) force pulls the apple and the moon toward the earth.

More than two centuries later, Einstein reinterpreted gravity again—in a way bearing more resemblance to Aristotle than to Newton: gravity is the natural way things move in the vicinity of matter!

For Aristotle, falling objects signified an earth at rest at the center of the cosmos. For Newton, they signified the attraction of all matter for all matter. For Einstein, they signified a non-Euclidean region of space-time. These are very different conceptions of the "same" natural phenomenon; one can scarcely point to the phenomenon itself without invoking a particular framework of explanation. While the quantitative difference between Einstein's and Newton's predictions were minute, the difference between Galileo's and Aristotle's conceptions had not, at the time, even been mathematically expressed. One can measure the acceleration of objects falling near the surface of the earth, and Galileo contrived an ingenious way to do so. But this could never have occurred to Aristotle, for whom the notion of the rate of change of motion with respect to *time* would have been meaningless, since for him motion was inherently relative to a special *place*.

To explain something scientifically is to create a context of possibilities in which to demonstrate why things are the way they are rather than some other way. It might be thought that scientific explanation does not itself need to be explained, because it is an activity that gives us satisfaction, like love or art.[10] However, love and art are human behaviors that we *do* seek to explain reductively. Similarly, one may legitimately seek to understand scientific explanation as an adaptive behavior.

The assumption that one natural fact can explain another is the basis of reductionism. This in turn, however, assumes a logical structure built into nature itself. That is a reasonable working assumption if one holds that logic generalizes empirical experience. But that is an assumption about the human mind as much as about nature. *We* do the generalizing, the deducing, and the explaining. On the other hand, if logic is held to be an absolute, independent of the human mind, one is at a loss to account for its correspondence with natural reality except through some metaphysical belief, such as that logical structure was imposed on nature by a creator god, or that nature is an unfolding of eternal logical truths. If consistency is desirable in naturalistic explanation, then logic itself must be explained naturalistically. Should we take for granted that other intelligent beings on distant planets will have discovered the same logic and laws of nature, explaining things in similar ways, when other intelligent beings on *this* planet have not bothered (so far as we know) to engage in scientific explanation at all?

Scientific Thought and Physical Reality

How *does* scientific thought as we know it relate to physical reality? It is certainly not a literal mirror or open window any more than ordinary cognition is. Yet there is a parallel in science to the naïve realism of ordinary cognition. This is the presumption that theory can provide a "precisely identical mathematical double of the universe."[11] This presumption is the gist of realism in physics as some have interpreted it. A quantity is real if its value can be predicted without altering the system of which it is a part;[12] and a theory is complete if it can predict the values of all such quantities. However, perfect predictability characterizes deductive systems, not real systems.

Two major precepts that guide scientific thought are mathematical elegance and agreement with experiment. Yet neither guarantees truth. Kepler's system[13]—of planetary orbits inscribed in geometric solids—for example, was just too tidy to be true; and Ptolemy's cumbersome system of epicycles was more accurate in its predictions than some modern theories. Most scientists embrace the idea, made explicit by Leibniz, that every logically meaningful question that can be asked must have a rational answer. That is rather a matter of faith, since what makes nature *real*, in our ordinary human eyes, is precisely that it offers no such guarantee.

Science *models* observed phenomena, according to prescribed goals. In contrast, one simply *experiences* the world, without concern for one's (unconscious) cognitive modeling. In scientific cognition the relation of model to world cannot be presumed as it is in ordinary experience. The whole point is to experimentally test the fit between the theoretical model and the observable phenomena.[14] This is not straightforward, however, since experiments yield their results in test situations that are already prescribed by theory. In effect—and rather circularly—the experiment is a physical realization of the model; one deductive system (the theoretical model) is embodied in another (the experimental or observational setup). Consequently, experiment can reveal dead ends but not necessarily what other avenues to explore.

Deductive systems can translate into each other because they are products of definition. This is the profound idea behind the computer, which is an abstract machine theoretically equivalent to any other machine. Faith that the physical universe is ultimately comprehensible, even computable, rests on the assumption that it is, for all practical purposes, a machine, computer or equivalent deductive system. For, we understand the universe through mathematical models, and it is easy enough to confuse the model with what it models, even to conclude that the universe is literally a "mathematical structure."[15]

A similar idealist faith feeds the contemporary notion that *information* is the essential substrate of nature.[16] However, the "it from bit" philosophy, as it is familiarly called, involves a bit of circularity: while physical reality is supposed to reduce to digital information, information is something abstracted from physical reality in the first place! Such circularity typically involves the sort of category mistake typical of mind-body dualism. One wonders how concepts—exemplified by math, programs, or information—are supposed to give rise to something material, such as atoms and forces. But then "atoms" and "forces" are also ambivalently concepts and real things.

Language permits us to make nouns of actions such as computing and processing information, as well as to reify traditional concepts such as energy and entropy. Through such legerdemain, computation or information becomes the ontological foundation of physical reality. The history of science will eventually tell whether this is a passing fad or a new paradigm (if, indeed, there is a difference). Either way, such a view involves two distinct claims. The first is that any physical system can be exactly described by some formalism. The second concerns the basic stuff of which reality is composed. Both trade on the notion that nature is discrete in principle at the bottom level, which does not strictly follow from nature's *observed* discrete aspects.

Reification and Ontology

Reification assures a causal basis for phenomena. For example, a force or substance is suggested to underlie the unsolved mysteries of "dark matter" and "dark energy," essentially by giving them names. Dark matter originally concerned anomalous rotational behavior of galaxies; dark energy concerns anomalies in the gravitational behavior of the universe at large. Einstein's "cosmological constant" was initially a mathematical fudge factor now taken seriously as a force driving the expansion of the universe. Positing such forces or entities may be the best strategy to account for observations, yet questioning the current interpretation of observations is a valid alternative.

Randomness and determinism are often considered to be properties of things themselves rather than pertaining to effects of the observer's relationship to them, knowledge of them, or interaction with them. Mathematics is often used to treat as quantities what are effectively operations. The very concept of infinity, for example, reifies a process of counting in order to treat it as a conventional number, upon which other operations can be performed.

Even the notion of cause is little more than shorthand for expectancies we infer from observed patterns. Of course, one wants to know which things and processes actually exist, how they work, and what they truly *are*. However, this desire is largely driven by the search for a reliable ground for prediction of possible future patterns to be encountered. One seeks information that yields a formula for action. This is the deep psychological basis of the notion of causality, intimated by Hume, who also observed that the more consequential the event, the more one seeks to read into it a coherent meaning.

Reification yields a plausible underlying structure in which to believe. It is based on the reasonable assumption that nature must *have* such a structure, quite apart from the structuring imposed by cognition. The problem is to identify nature's real structures, given the inherent ambiguities of cognition and the need for certainty the organism brings to the table. However, if this structure could be revealed *unambiguously*, there would be no need for the organism to favor certain ways of organizing its cognition, or to have any choice or flexibility in how it perceives the world, which would straightforwardly be dictated by the world itself.[17] Theory trades on the realist belief that the world exists a certain definite way (counterfactual definiteness) regardless of how we cognize it. But this is not the actual situation facing the subject, for whom the world remains ambiguous and open to multiple interpretations. Science gets by through assuming realism where it can, and by carrying on anyway where it cannot.

Physical concepts and laws are defined in terms of measurable quantities. Presupposing an *entity* to carry the properties measured involves a metaphysical leap. It involves circular reasoning when the entity cannot be verified other than through those measurable quantities, as famously illustrated in the mutual definitions of force and mass in Newton's formula, $f=ma$.

Though it is the very basis of scientific materialism, the notion of *material substance*, or matter—whose primary measurable property is mass—has a tortuous history. Even in classical physics, any measurement of a "quantity of matter" involves an exchange of "energy," in an interaction with some measuring system.[18] While this may only negligibly affect the measurement, it muddies the notion of substance by appealing to an even more elusive concept, energy. In turn, that concept had first to be dissociated from matter (which had once served as its substrate) before it could be (re)unified with the concept of mass. When we think of the equivalence of mass and energy we are disposed to think of one substance transformed into another. However, both mass and energy are measures, not substances. We are only ever dealing with measures, not with what is measured. Yet the concept of energy shifted from being a measurable *state* of matter to being an ontological *entity* in its own right—something that itself *has* mass as a property or state.

Thus, both mass and energy are manifestations of something putatively substantial, yet defined through operations that are relational. Indeed, Dalton's "atoms" had not been real physical particles but a mere accounting device to keep track of proportions of elementary materials.[19] We now give a common name (energy) to various phenomena revealed in distinct situations by different instruments, as though a definite single thing is involved, whereas it is really a common purpose involved.[20] One might wonder whether a society uninterested in motors could have conceived a unified concept of energy, let alone of entropy (a concept that evolved somewhat on the analogy of energy). Entropy is a property of the whole system; however, it is misleading to speak of the entropy of specific parts, or of a *flow* of entropy from one part to another, as though it were a sort of substance.

Similarly, we give a common name (mass) to a measurable property that manifests in two distinct ways and may have more than one origin. Modern theory has the mass of the proton, for example, originate in the internal energy associated with its constituent quarks, but must account for the minimal mass of the quark itself and the electron differently.[21]

The notion of the *field* also involves reification. The strength of magnetic attraction or repulsion, for instance, can be measured through interaction with other magnets or electric charges. The surrounding space can be mapped in terms of the potential for such interaction at each location, and in that sense the magnetic field is a mathematical device. But the field has come to be regarded as a real entity permeating space. While magnetism is palpably real, does it add to our knowledge to think of it as substantial in the way that matter seems to be? Certainly, the field concept has proven extraordinarily useful. Moreover, the electric field, for instance, can produce effects associated with mass, which is traditionally identified with substance. On the other hand, imagining the material reality and mechanical properties of the electromagnetic field led to a dead end in nineteenth-century physics. Yet, this was hardly the end of the story, since the concept of "vacuum energy" again suggests a kind of substantial aether.[22]

In a sense, even causation is reified—in the concept of force, which no doubt arose from personal muscular exertion, extended to the impersonal influence one external thing is seen to exert upon another. Force represents efficient cause defined mathematically. Yet abstraction, and a common name, may give a misleading impression of a monolithic notion. Why, for example, are there four "fundamental forces," and what exactly do they have in common? Why are there such huge differences in their range, strength, and sign?[23] These, of course, are questions inspiring (and plaguing) modern theories of unification. Yet unity of ideas does not necessarily imply unity within nature. We could be misled by nothing deeper than that one word is used to cover several phenomena.

People and individual animals seem to be discrete and autonomous things. But even these apparent individuals are ill defined at a finer level, consisting of multiple organisms including parasites, bacteria, and viruses, enmeshed with the community of cells that belong genetically to the organism and far outnumbering them. The cell may be the paradigm of the discrete organism, the equivalent of the particle in physics, but neither is necessarily fundamental.

There are simply no absolutely discrete or permanent individuals in the wild. Trees, for example, are not distinct freestanding entities. They are always changing; their roots are entangled with those of other plants and with various small creatures and microorganisms, their trunks and branches entwined with vines, mosses, epiphytes, rocks, etc. Even a boulder is not a well-defined object, since it is friable, in process of disintegrating, and often covered with mosses and lichens that hasten its changes. Much less is a mountain or a cloud a clearly definable thing.

In contrast, mathematical idealizations *are* well defined—and eternal. The problem is that they do not literally exist in nature. They are concepts useful for human purposes, just as the ordinary perception of *objects* is useful in daily life. Mathematical objects exist by fiat—as products of definition—yet they are based on experience of real things. This is so, on another level, even of the perceived objects of daily life insofar as they are constructed through cognitive processing, in language, or literally through human craftsmanship.[24] While words name things that seem real and integral, they also lend reality and integrity to the things they name. We are increasingly surrounded by human artifacts—products both of definition and of industry—which in turn furnish an expanding paradigm for the very notion of "object."

Thus science capitalizes on a built-in aspect of perception and thought. For evolutionary reasons, it is essential to us to perceive the world in terms of discrete things. A basic mathematical intuition is nevertheless to treat the world as a continuum. Here the needs of an embodied consciousness compete with its capacity for logical abstraction. The conflict between these inclinations leads to logical paradox and mathematical problems of infinities and discontinuities: the old problem of the discrete and the continuous.

Like religion, science provides a common framework of practices and principles that facilitate agreement about what exists. The ontological basis of the scientific framework is the objective presence and continuity of the natural world we share in common, as opposed to idiosyncratic perceptions or the beliefs of a particular community. In principle, modern science relies upon an agreement that nature, rather than some doctrine, shall be the ultimate arbiter of truth. However, scientific theory *is* a doctrine. Belief enters here too, for the above-mentioned agreement already involves faith in a shared understanding. If the history of science has taught us anything, it is that ideas about what exists, and the very concept of nature, are continually up for revision.

"Theory" and "theology," in fact, share an etymological root grounded in a common preoccupation with what ultimately exists. Like scientific theory, theology tends to focus on the object rather than the subject, on the niceties of the spiritual world. But there is more to religion than theology, just as there is more to science than theory. Unlike science, which dwells on the niceties of the material world, in religion the subject is also a focus. The best of religion speaks to the subject, encouraging an attitude of receptiveness quite different than the value of control or positive knowledge prevalent in science.

The very directedness of mind makes for a transitive relationship to experience. One knows or believes *something*, one has faith *in* something definite, whether a theory or a theological doctrine. It hardly computes in the rational mind simply to *have faith*, without an object that serves as rationale to justify it. Hence, dogma—whether religious or scientific—is obliged to spell out in detail the "reality" in which one is to believe. The question remains, can an attitude of inquiry be cultivated that suspends the need for dogma concerning definite realities? It is a question that applies as much to science as to religion.

Why is Nature Intelligible?

Should one marvel that the world is comprehensible? The intelligibility of nature was taken for granted in an era when it was believed to be the divine creation of a rational Mind. Yet nature's intelligibility is no more surprising if one believes that we are natural creatures whose embodied minds have been attuned to reality by nature itself.[25] What *is* surprising is that nature often appears to conform to simple models, based on limited experience, even in areas far outside that experience.

Most basic physical notions derive from familiar encounters with things and processes on the human scale: the behavior of ordinary objects at ambient temperature and pressure and over human timelines. The behavior of matter outside this range often proves counterintuitive. Extremely fast, small, or highly complex things defy ordinary expectations as may things at extreme temperatures, densities, and pressures, or on the largest scale. One must be prepared to subordinate intuition to experimental fact in these domains, since intuition did not grow up in them.

A number of developments in the last century conspire to undermine the expectations of common sense: the revelations of relativity and quantum theory; the new disciplines of complexity and chaos, and of high-energy and low-temperature physics; Gödel's defeat of the nineteenth-century program to axiomatize mathematics; new understandings in biology and ecology. These developments also suggest limits of theoretical programs that rely on simple idealizations, isolated systems, and the separation of subject and object. They hint at nature's real complexity and interconnectedness, and at the scientist's participation (which often consists in dividing things up artificially), as well as the participation of all things in all other things. Modern concepts such as entanglement and non-locality point to an essential wholeness even in the nonliving world. While categories of thought, even logic itself, are based on experience in everyday life, the quantum world defies many such intuitions. Yet its challenges point to the very reality of nature as a presence transcending our limiting ideas. They remind us that understanding is always a tentative affair, made from *within* nature, not from a godlike perspective outside it.

While it appears that the world is knowable (and livable!) because nature obeys rational principles and hierarchical laws, one may wonder to what extent these simply express human preferences and thought patterns, projected onto a nature that generously produced scientists along with the things they study. Of course, the answer must include both. The world is obviously suitable for life, since here we are; it is intelligible for the same reason. It is knowable because we do create knowledge that works for us, within a context that allows our being. Yet one may wonder whether scientific knowledge will continue to allow our being in the longer run. In any case, it is important to sort out what is intrinsic to the world from our scientific projections upon it. In one sense, science claims to have already done this long ago, simply by eliminating secondary qualities and subjective factors in its method, as originally prescribed by the Enlightenment thinkers. One may also wonder, however, whether there are still subtler ways in which science remains embedded in its own cognitive biases.

The "principle of sufficient reason," credited to Leibniz, might reflect such a bias. It suggests that there should be an answer to every reasonable question about why the universe is as it is. Yet this faith may be no more than wishful thinking, whose success as an evolutionary strategy in ordinary realms bears no absolute guarantees, unless one simply accepts (as Leibniz proposed) a divinely preestablished harmony between reality and thought. Similarly, the principle of the "identity of indiscernibles" depends on two things sharing all possible relationships or properties, for their distinctness hinges on having distinct relationships to the rest of the world and thus distinct properties. Whether things can be held to be identical on that basis depends on the possibility to exhaustively enumerate a finite list of those properties or relationships. But such a complete list is possible only for a deductive system, where the properties and relationships are definitional; the principle already assumes that the world is such a system.

Measurement is theory-dependent insofar as it presumes quantities that can be isolated as *the* pertinent variables of the theory, and that faithfully represent properties of the world rather than of the investigating apparatus and method. To what extent can reality be separated from the interactions of our instruments with it? How well can variables of interest be distinguished from "noise"—that is, from information that is already presumed irrelevant? Perhaps the variables of a theory are *not* exclusive nor exhaustive and do not pick out clearly identifiable properties. *Defining* them mathematically, however, gives the illusion of completeness and definiteness, masking the ambiguities involved in the experimental or observational setup, which include shielding, interpretation of signals, statistical spreads, etc.[26]

Most working scientists do not have time to concern themselves with questions such as why the world is comprehensible. Einstein was an exception, a philosopher-scientist who marveled—with Kant—that "the eternal mystery of the world is its comprehensibility." Einstein had a lifelong concern with the relationship between thought and reality, between "sense impressions" and "concepts," acknowledging that there is no *a priori* connection between them. Yet, while concepts are free constructions, they cannot be arbitrary, since they refer to a real natural order. If Einstein finds that order wondrous, it is against his expectation of *chaos as the natural state of things*. Yet we have no a priori justification for any expectation whatsoever—whether of order or of chaos! Einstein's sense of science as a heroic quest—from outside nature, for the right concepts that correspond by intellectual force to the real order within it—must be understood in the light of this expectation. His "classical" position is understandable, since it predates the understandings of evolutionary epistemology. Such an expectation does not take into account the fact of belonging within the natural order, which shapes our categories of thought and very expectations, and is responsible for the "preestablished harmony" between ordinary cognition and natural reality. It is this belonging that must ultimately also be responsible for the harmony possible between scientific thought and natural reality.

A Few Loose Ends

I believe there will always remain unresolved scientific questions, simply because nature is *real*. Underlying these, there remain also perennial *conceptual* issues, which implicate not only the nature of the object but of the knowing subject as well. Approaches are sometimes inconsistent, reflecting deep conundrums that are perhaps irresolvable and which result in methodological difficulties like "renormalization."[27] We still do not know whether the universe is finite or infinite or whether there is a "bottom" to the complexity of matter. Nor do we have a definitive ontology telling us what the physical world fundamentally *is*. Such questions cannot be separated from our own mental organization.

It seems the world could not exist without discreteness on some basis.[28] Yet, discernible parts, even if considered irreducibly fundamental, might be no more than partitions of an energy continuum. Both continuity and discreteness are intuitive notions that reflect common experience. One naturally wonders of what parts anything is composed; yet a part with true integrity must be internally structureless or empty—a plenum or a void. If it does not consist of parts, however, it cannot be understood reductively, and so must be taken on faith as an ultimate unit. This is equivalent to positing it axiomatically as an element of a deductive system. While such particles might act upon each other, either through direct contact or across an intervening space or medium, one is at liberty to further question both the indivisibility of particles and the composition of the space between them and within them, posing the problem on a deeper level. One is at once committed to understand a continuum in terms of atoms, and atomism in terms of a continuum.

Classical physics, for the most part, simply ignored such issues. We have acquired the habit of relegating nature's weirdness to the quantum realm, while in fact similar conundrums trouble the classical realm as well. They ultimately stem from limits of reason and cognition, pointing to our embodied origins. We see and reason as we do, not because reason and cognition are uniformly true to reality but simply because doing so has favored our survival thus far, with or without logical consistency.

Perfectly elastic collision between theoretically integral objects is an idealization that contradicts the requirement of causal continuity, since an instantaneous transmission of force is implied.[29] To be continuous, the transmission of force presumably must take time, acting across distance or transmitted through a medium. And yet one may ask what it is exactly that takes time in such a process. How can elasticity itself be understood, except in terms of smaller parts acting at a distance or in terms of smaller elastic objects in contact, which simply regresses the problem?

The numerical separateness of physical things rests conceptually on their impenetrability. Two ordinary solid objects cannot occupy the same volume at once; otherwise, they could not be counted distinct on the basis of spatial separation. (Waves, on the other hand, can interpenetrate but have no clear or lasting identity.) However, impenetrability is relative, since it depends on forces that may be overcome in extreme conditions, for example, in degenerate matter or unification of forces at high energy.

While nature cannot be reduced to a deductive system, it does often seem to *mimic* one. Many classical systems, such as the solar system, behave like their mathematical descriptions to great accuracy.[30] Quantum entities in particular appear to manifest the precise and simple integrity characterizing products of definition. How to separate the fundamental particle's intrinsic discreteness (if such exists) from the discreteness it enjoys merely as a theoretical product of definition? Since a black hole is considered internally structureless, should it be regarded as an elementary particle? Or perhaps an antiparticle, since it is the opposite of impenetrable?

Clearly, the extremes of scale do not play by the same rules as the human-scale world. Yet, there is no absolute divide between the physics of different scales, and quantum phenomena are demonstrated in laboratory for increasingly large objects. If one thinks in classical terms about the Pauli exclusion principle, for example, it appears that an electron "knows" the states of its fellows through no conceivable causal process. Similarly, to ask the *cause* of quantum fluctuations is to insist on a deeper level of determinist description, to which yet deeper limits might again be found—ad nauseam. For the most part, nonlocal effects in the quantum realm do not make sense in terms of ordinary experience. Yet, nonlocality was a problem long before quantum theory. Gravitation and magnetism had early suggested "occult" forces acting mysteriously and instantaneously across distances.

Einstein offered no dynamic (causal) explanation for the absolute speed limit, c. Rather, he set it forth as a postulate, along with the invariance of laws in inertial frames. While necessary, this cutting of the Gordian knot violated common sense. The problem of a medium of transmission was swept under the rug. Special relativity (SR) could be interpreted either epistemologically or ontologically (the first part of Einstein's original paper was "kinematical," while the second part was "electrodynamical"). Lorentz, Fitzgerald, and Poincaré all seem to have taken the length contraction quite physically as a dynamical effect of forces differing in a rod in motion and in one at rest. Einstein held rather that the length contraction is a necessary consequence of his two postulates.[31] But what does it mean for a *physical* contraction to be caused by the *theorist's* postulation? This seems oddly reminiscent of mind-body dualism.

The arguments of SR were first presented in terms of mutual line-of-sight effects, involving light signals between distant frames of reference, in uniform motion away from or toward each other. The contraction effect is *mutually* perceived, so cannot have an ontological sense based on agreement about which observer's measuring stick has "really" shrunk. On the other hand, there *is* objective evidence for the reality of time dilation, whatever its explanation. Moving clocks can be brought back together to compare the time elapsed. But no one has proposed that a travelling twin returns to find herself shrunken or compressed! If time is merely one of four qualitatively similar "dimensions," how do we account conceptually for this difference between "space" and "time" contractions?

It is ironic that general relativity (GR)—as well as quantum theory—proposes "empty" space as a medium of sorts, which special relativity avoids. Physics is still encumbered with dual approaches to gravitation and its fit with the other basic forces. Following Newton's example, Einstein offered no explanation of why matter affects the space around it. GR explains gravity in terms of geometry and the continuity associated with waves and fields, while quantum theory explains it as an exchange of particles, which can mysteriously arise from the vacuum itself. At the heart of physics, preventing its unification, there is still a dualism of wave and particle, of continuum versus discrete.

Perhaps the ultimate logistical problem concerns the origin of the cosmos: how something could come into being from apparent nothing—if indeed the big bang can be called an origin. The paradigm we have for creation ex nihilo is religious (and linguistic): a conscious agent simply *declares* the world into being. Modern cosmologists follow this example, simply declaring model universes into being by specifying various parameters of the equations involved. The acts of *physical* creation with which we are familiar (human artifacts) involve form imposed on a preexisting material, which is found not made. Whatever physical creation process is theorized, perhaps there will always be a regressive problem to explain the aboriginal material—the "waters of the deep."

If the universe had a beginning, one might well wonder what preceded it. If it is finite, one might wonder what surrounds it. Modern cosmology has sophisticated answers to such questions—for example, that time and space themselves came into being with the big bang and that space (and time?) can expand yet be "closed" so that there is no boundary. On the other hand, some cosmologists propose a time before the beginning, in which the universe we know arises as a bubble in a vaster froth—a space containing space and a time containing time. In that case, boundaries of time or space are simply deferred to a larger context. Either way, physicists trust far more to abstruse mathematics than to common sense. Their answers can be challenging for the rest of us.

Common sense is hamstrung by experience on the human scale, where things normally have limited size and duration, effects have causes, and things either exist or do not, in one place rather than another. Reason generalizes common experience, in the broadest strokes, but also extrapolates (sometimes wildly) from it. Our experience is of finite things, yet we conceive of infinity and even of various "sizes" of infinity, according to formalized rules of reasoning. Yet reason comes up against experience in contradictions that arise when its extrapolations are unreliable or when nature itself does not appear to be as reasonable as we would like. In any case, it seems clear that we should not forget limitations of human thought when considering inconsistencies in science or apparent contradictions in nature.

Notes

1. Cartwright, Nancy. 1983. *How the Laws of Nature Lie*. Oxford: Oxford University Press, p. 19.
2. McKibben, Bill. 1989. *The End of Nature*. New York: Random House, p. 58.
3. Faun, Feral. "Nature as Spectacle: The Image of Wilderness vs. Wildness." Anarchist Library. http://theanarchistlibrary.org/library/feral-faun-essays#toc3.
4. *Autopoiesis* is a term coined by theoretical biologists Humberto Maturana and Francisco Varela.
5. Deason, Gary B. 1986. "Reformation Theology and the Mechanistic Conception of Nature." In *God and Nature: Historical Essays on the Encounter between Christianity and Science*, edited by David C. Lindberg and Ronald L. Numbers. Berkeley: University of California Press, p. 177.
6. While Aristotle did not literally use the expression "monsters of nature," he did speak of monstrosities, for example what we now call genetic deformities. The sense of the term here used is to underline his resistance to the notion of experiment as an external interference which (by his definition) could not reveal the true natures of things. The modern version of this idea is what philosopher Nancy Cartwright calls a *nomological machine*.
7. Osler, Margaret. 1996. "From Immanent Natures to Nature as Artifice: The Reinterpretation of Final Causes in Seventeenth Century Natural Philosophy." *The Monist* 79:388–407.
8. Smolin, Lee. 2013. *Time Reborn: From the Crisis in Physics to the Future of the Universe*. Toronto: Alfred A. Knopf Canada, pp. 148–49.
9. Weinberg, Steven. 1992. *Dreams of a Final Theory*. New York: Pantheon, p. 219.
10. Ibid, p. 26.
11. Smolin, *Time Reborn*, p. 251.
12. Powers, Jonathan. 1982. *Philosophy and the New Physics* London: Methuen, p. 147.
13. http://en.wikipedia.org/wiki/Mysterium_Cosmographicum.
14. In the case of ordinary cognition, this has presumably already been tested on an evolutionary time scale, such that those organisms exist that have a "correct" relationship between their cognitive models and the external world—that is, a relationship that permits their survival.
15. Tegmark, Max. 2007. "The Mathematical Universe." http://arxiv.org/abs/0704.0646.

16. Alternatively known as *pancomputationalism*.
17. James W. McAllister, James W. 2011. "What Do Patterns in Empirical Data Tell Us about the Structure of the World?" *Synthese* 182:73–87.
18. Jammer, Max. 1997. *Concepts of Mass in Classical and Modern Physics* Mineola, NY: Dover, p. 4.
19. von Baeyer, Hans Christian. 1999. *Warmth Disperses and Time Passes: The History of Heat.* New York: Modern Library, p. 10.
20. Bridgman, Percy Williams, 1941. *The Nature of Thermodynamics.* Cambridge, MA: Harvard University Press, p. 114.
21. Frank Wilzcek, Frank. 2008. *The Lightness of Being: Mass, Ether, and the Unification of Forces.* New York: Basic Books, p. 200.
22. Cf. Laughlin, Robert B. 2005. *A Different Universe: Reinventing Physics from the Bottom Down.* New York: Basic Books, p. 121.
23. Unzicker, Alexander, and Sheilla Jones. 2013. *Bankrupting Physics: How Today's Top Scientists Are Gambling Away Their Credibility.* London: Palgrave Macmillan, p. 205.
24. Areas of the human cortex involved in language and in toolmaking are suggestively proximal. Their common denominator may be *sequencing* (of words or of actions). See the 2010 PBS series *The Human Spark.* www.pbs.org/wnet/humanspark/.
25. By the same token, natural beauty is no surprise if our esthetic sense is likewise shaped by evolution.
26. Cartwright, Nancy. 1999. *The Dappled World: A Study of the Boundaries of Science,* Cambridge, UK: Cambridge University Press, p. 152.
27. Wikipedia. "Renormalization: Any of a collection of techniques used to treat infinities arising in calculated quantities."
28. Without it, the electrons of Rutherford's classical atom, for example, would continuously radiate away all their orbital energy and fall into the nucleus.
29. Jammer, Max. 1957. *Concepts of Force.* Cambridge, MA: Harvard University Press, p. 211.
30. The question of accuracy itself begs an absolute reference that does not exist. At best, we can subjectively evaluate standards and compare different measures, for example, to compare siderial time with atomic time.
31. Pais, Abraham. 1982. *Subtle Is the Lord: The Life and Science of Albert Einstein.* Oxford: Oxford University Press, p. 141.

3

What Is Science?

"The eternal mystery of the world is its comprehensibility."
—Immanuel Kant

"The problem with experts is that they do not know what they do not know."
—Nicholas Taleb[1]

Introduction

In our Western traditions, scientific knowledge has come to involve taking things apart in thought, if not literally, as a way to investigate their normal integrity. Analysis is a phase of a dialectical process in which things are put back together in a more encompassing and empowering synthesis: a view of the world as unified, predictable, and "natural."

Prescientific traditions, which we now call myth, gave accounts of the way things normally are and how they got to be so. Further explanation usually was only required for *exceptional* events, in their unique detail: why a *particular* thing befell a particular person at a particular time. Often the explanation was in terms of benevolent or malevolent intentions, and a culprit was sought for some breach in the proper order. By contrast, in their search for the origins of things, the literate ancient Greeks sought general principles of an impersonal kind, a baseline of constancy underlying change. Whereas myth often served social cohesion, Greek thought was individualistic, contentious, competitive, and skeptical. Opinion was argued with persuasive tools of rhetoric. Whether or not nature was to be taken apart in thought, the arguments of others certainly were. While preliterate cultures tended to accept multiple versions of oral narratives,[2] the modern mentality, descended from the Greeks, tends to insist rather on a singular, correct, and *textual* account. The creation stories of oral traditions are sanctioned belief that does not necessarily lay claim to a unique, objective, or absolute truth, which is rather a modern idea.

In part, the coherence of the *scientific* creation story is a product of revisionism among educators, science writers, historians of science, and scientists themselves.[3] This reflects the demand for a singular true, objective and potentially definitive account. Yet science requires ongoing empirical investigation, reasoned argument, and debate in order to achieve this. Paradoxically, the desired narrative is supposed to stand outside time, yet is continually revised. To dwell on conceptual history, or on changing epistemic standards, can undermine the requirement for a definitive account. While the tortuous story of how this account is worked out is deemed irrelevant to its eventual truth, this story reflects the actual scientific process. As a consequence, there is both a public and a private scientific narrative.[4]

One of the frustrations of learning science is the tidy revisionism of textbooks, where the current scientific view is presented almost axiomatically, as a definitive account, not as a record of the heroic and often fruitless struggles that led to it. Reigning ideas are presented as fact, with little mention of prevailing biases, disqualified competitors, or minority opinions. This gives a false view of the process of science and its cumulative progress. It also gives the impression that nature necessarily corresponds to current theories.

By definition, objective truth cannot be merely a product of human history or cultural circumstance. Scientific truth must stand on its own—or, rather, on nature's—testimony, beyond time, place, and human fallibility. The authority of science ideally rests on empirical evidence and logical merit, not on its social value or practical utility or on the force of rhetoric. It is no coincidence, however, that religion makes parallel claims to a singular transcendent truth. For the scientific ideal of truth descends from the Christian ideal of revealed wisdom as well as the Greek ideal of deduction and skeptical debate. While modern intellectual tradition views religion as a cultural and historical product, believers rarely view their beliefs in that light. Science, too, tends to exempt itself from self-examination, which is left to historians and philosophers of science. Yet reflection on the methods of knowledge had been an integral part of natural philosophy and may be again.[5] We shall see that the Scientific Revolution served above all to redefine natural philosophy as something whose exclusive focus was the external world, not its own process.

Understanding and explanation can be understood in objective or subjective terms: either in terms of *causes* (such as may occur between inert things) or in terms of *reasons* (such as people may offer as justification for their actions or ideas). Science has nominally abandoned the sort of *why* questions that appeal to reasons, in favor of descriptions of what naturally occurs and how, in terms of efficient causes. It does this at the price of limiting discourse to third-person description within defined bounds, beyond which the question of *why* is taken up by religion or philosophy. Yet the boundaries of legitimate science are continually expanding with new theory and observational capabilities. Scientists themselves are often not content to be confined within what is strictly deemed scientific explanation in their era or specialty. The heroic role of science is to assimilate the unknown to the known, which means to human thought systems and to specific terms of reference. It is at this frontier that cause and reason merge. The attempt to rationally understand nature is ultimately an attempt to assimilate the found to the made: to give *reasons* for apparent natural patterns; to make them comprehensible in terms of human intentions, conventions, and inventions. This is the larger significance of mathematical *models* in the scientific narrative, which transcribe natural reality into defined, rational terms. It is the very basis of *theory*, which derives from the same root as *theology*, and which answers the question: if you were God, how would *you* have made the world?

The Scientific Revolution

Science today generally means reliable systematic knowledge—particularly about the natural world. Reason underpins its process, and philosophical materialism underpins its content. Thus, the scientific method is the bona fide path to knowledge, while matter and energy are fundamental things held to exist. However, the word *scientist* came into usage only in the nineteenth century. Before that, those who studied nature were known as natural philosophers. Philosophy, of course, had included diverse forms of knowledge and approaches to it, the whole range of human experience, and a variety of beliefs about what fundamentally exists. It also included second-order ideas *about* knowledge and the proper path to it.

The distinctly modern method of acquiring knowledge of nature began with the Scientific Revolution. This coincided more or less with the Protestant Reformation. A new conception of the inertness of matter paralleled the Protestant insistence on the weakness of sinful flesh, which implied salvation by faith alone. Both underlined divine power over nature and mankind. Just as the Reformation rejected the metaphysical speculations of Catholic scholasticism in favor of a more literal reading of the Bible, the Scientific Revolution abandoned the wide-ranging concerns of philosophy, to focus on the physical world apart from subjective concerns. Facts were preferred to speculation and interpretation. In the case of science, this meant abandoning the more self-reflecting aspects of philosophy as well. Above all, it meant that the Scientific Revolution redefined natural philosophy as what we today call "first-order" science: study of the external world in exclusively objectivist terms, without self-reference and amenable to mathematical treatment.

Aristotle is recognized as the great natural philosopher of antiquity. Though a keen observer, he conceptualized nature in ways that were rejected in the Scientific Revolution as old fashioned and unfruitful; often he asserted little more than that things behave as they do because it is their "nature" to do so. According to Aristotle, all movement is either "natural" or enforced by some agent. A body moves in its own characteristic way, toward its appointed natural place, unless forced to deviate by another body. In the course of its movement, it may cause another body to deviate from its natural motion. The phenomenon of weight was but an aspect of natural motion: things moving to their proper place. Thus, it could not be used as a standard measure of *force*, which implied a deviation, and this impeded the development of modern dynamics.[6]

Aristotle thought of change more in terms of place than time. His broad ideas about causality were later narrowed to include only *efficient cause*, which in his system represented a breach in the normal order of things. In contrast, the new science of dynamics became the general analysis of change over time: efficient cause *became* the normal order. Aristotle objected to constraining nature in artificial ways through experiment. This was overruled in the new science and experiment became standard procedure. It isolates efficient causes expressly to see how nature responds to "breaches" brought about by "outside" intervention. Though we've come to think of experiment as a way of *observing* nature, Aristotle had shrewdly cautioned that it *interferes* with nature, producing only artifacts. While modernity does not hold with Aristotle—that the parts of nature may exist *for the sake of* greater wholes—nevertheless we must recognize that parts cannot exist without reference to them.

Copernicus realized that gravity was aimed not just toward the center of a spherical universe, nor toward the center of the earth, which had been supposed to coincide with it, but was a tendency of *all* matter to coalesce into spherical form, like raindrops. This could only be so if it acted equally in all directions. Nevertheless, under Aristotle's persistent sway, Copernicus continued to view gravity as resembling desire, assigning to it a final cause: to achieve the perfection belonging naturally to the sphere. From his planetary observations, Kepler realized that the attractive force between celestial objects falls off with distance between them and so concluded that the force is physical rather than the influence of some "motive intelligence."[7]

Galileo held mathematics to be essential to the description of nature, proposing to replace the nebulous "why" of religion and philosophy with the precise "how" of mathematical description. He proclaims the axiomatic method as the opposite of poetry, yet ironically relies on what is ultimately a poetic allusion: nature as geometry.[8] In particular, time could be represented geometrically as a spatial dimension. Time analysis reflects the shift, from Aristotle's static system concerned with place, to the new science of dynamics.

If Galileo was the first great scientist, Francis Bacon was the great advocate and apologist of the new science, who shaped its present form and won its political acceptance. Bacon's program was the pursuit of useful knowledge and practical technology rather than "serpentine" metaphysical speculation. The context of this program, however, was the Christian project to recover Adam's original perfect knowledge, lost at the biblical fall. Like others of his era, Bacon regarded Aristotle as an impediment to this recovery.[9] For Bacon, the "words" of the language of nature are not mathematical but are the natural things themselves. However, nature contained no preordained lexicon, and Bacon was well aware that what united spoken languages was their structure more than their vocabulary. This approaches the understanding of Galileo: mathematics represents the grammar, if not the vocabulary, of the natural world.[10] Even so, Bacon cautioned that mathematics should not dominate physics.[11]

Like Galileo, Bacon objects to the hubris of scholasticism, but substitutes for it another pretension: to know nature "immediately." He prefers to let nature speak for itself, downplaying the role of speculation in favor of a more direct transcription of natural fact. Bacon's attitude is consistent with the Protestant approach to biblical study and to life. But he has a particular method of active investigation in mind, through which both the pitfalls of mere sensory knowledge and those of metaphysical speculation can be avoided: the scientific method of controlled experimentation. The creative mind is thereby disciplined to find clever ways to "interrogate" the natural world. Natural inquiry is redefined, from passive observation to actively prying answers from nature. Bacon was a lawyer who worked for the Crown prosecution; the terms in which he advocates forcing nature to divulge its secrets mirror the proceedings of the Inquisition. Nature was put on trial, "tortured" through experiment.

Mysticism and natural magic played as large a role as mathematics in the age, as reflected in the multiple interests of Newton. The dream of a new philosophy inspired alchemists such as Paracelsus and Fludd, as well as rationalists like Bacon, Descartes, and Galileo. With Descartes, however, there is a clear separation of physical from mental terms in the description of nature. His program, accordingly, seeks a purely extensional description, avoiding the concept of force, for example, which he felt still smacked of the tactile senses. Consequently, his theory of gravitation shunned action at a distance (an "occult" force), attempting a kind of mechanical field theory. Huygens, Newton, and Hooke also toyed with mechanical models to explain gravity, and Newton himself disclaimed the notion of action at a distance that became associated with him. While he generally thought science should be separate from religion, this did not prevent him from privately interpreting gravitation as the direct action of divine will.

The Metaphor of Jurisprudence

One may compare scientific method with jurisprudence, insofar as both follow regulated procedures for the collection and weighing of evidence. In science, however, nature is both defendant and witness. As theories can be overthrown, but never absolutely proven, the testimony of experiment or observation can tell us only whether the evidence contradicts the specific "accusation." Otherwise, it tells little about the party on trial. The fact that nature may be absolved of given claims does not tell us much about her or what other claims might have been more appropriate. Moreover, it is misleading to figure nature as witness in scientific procedures when it is actually human researchers who give testimony.

Science and jurisprudence both investigate apparent breaches of "law." It is significant that the same word denotes two entirely distinct concepts. Laws in the juridical sense are made by human beings, when not by gods. Either way, they are matters of decree. Scientific laws are formulated by scientists, but are supposed to accurately express patterns *observed* in nature, not decreed. One kind of law is made, the other found. Conflating the two leads to the belief that laws of physics have some regulatory or governing power. Though it is passable as a figure of speech to say that nature *obeys* the laws of physics, it is tempting to take such expressions too literally. While society's laws are known and followed because they are legislated, in science the purpose of inquiry is to discover what regularities best describe nature.

Jurisprudence evolved protocols to regulate testimony, requiring witnesses to bear responsibility for their claims. Science too has protocols to evaluate scientific testimony—whether of nature or researchers. Jurisprudence has a practical focus, with interest in general philosophical arguments about the meaning of evidence or testimony only when it bears on the validity of specific claims or how they should be interpreted in a specific case. Similarly, the court of scientific opinion is most interested in philosophical argument when it makes a difference that is decidable by experiment or applicable in technology.

Science could only develop in an apparently law-abiding universe—and in a law-abiding society. The meaning and rationale of natural law, however, remain matters of faith that can only be supported through some form of circular argument. It can be no coincidence that science was fostered in a society whose faith was placed in a rational God. The Royal Society was founded in the wake of the Restoration, following the chaos of religious wars, by men anxious to put aside doctrinal bickering in favor of order and truths they could agree upon. This may shed light on why explaining the origin of order in the cosmos is psychologically challenging for naturalistic theories that rely on chance, such as Darwinism and quantum theory.[12] It may be one reason why science has taken such refuge in mathematics—the sheer embodiment of reason and order—as the essence of scientific law and even as the ontological basis of nature. The glaring alternatives are theocracy, on the one hand, and a fundamentally inexplicable universe on the other.

Cognition, Narrative, and Theme

As a form of cognition, science focuses on what it presumes to exist independent of cognition. The epistemic circumstance of the scientist mimics that of the brain, sealed inside the skull. Both situations require inference. Just as the brain relies on the input of receptors to make inferences about the real world, so the scientist relies on instruments' readings. Instrumentation extends the senses, substitutes for them, and also objectifies the properties to which it responds.

The outward-looking brain routinely projects its own processing as events happening in real space. Such a projective faculty evolved for an obvious reason: the need to take the external world seriously, as a place full of consequence for the organism, requiring decisive action. In both ordinary cognition and in science, therefore, there is a tendency to objectify relationships and to render the ambiguous definite. In the context of science, this means interpreting data in terms of entities, forces, or laws. It also means redefining natural phenomena as theoretical constructs and treating such constructs as realities.

While science may be conceived as a cognitive activity, it may also be likened to a social game, with rules and premises. Discovery has an unpredictable element of input corresponding to the roll of dice. Interpretation builds on cultural assumptions and on motivations arising from embodiment in a larger evolutionary "game." Naturally ambiguous things are transposed as precision elements of the game—idealized conceptual objects, such as billiard balls and "particles." Perception and science fulfill similar needs, pursuing cognitive strategies on different levels and scales. Each provides leverage rather than a transparent window on the world.

The game tree of science, like the tree of life, develops irreversibly and somewhat haphazardly. Progress means building on certain premises, then on modifications of these, and so on. While the game satisfies rules at a particular time, the rules evolve. To the extent the game is a closed and fixed system at any moment, *science* is deterministic whether or not nature is. One may view it as the best system we have for dealing with nature. While all systems are finite and limited, according to philosopher David Lewis the "best system" strikes the optimum balance between simplicity and strength—a notion intuitively compatible with the strengths and limitations of embodied creatures, which seek the optimum use of their resources. This does not mean that the world has a unique optimum representation. Our representations may simply reflect our particular embodiment, while a different embodiment or environment might impose different constraints.

Science has a second role, with a corresponding social obligation: to fulfill the need provided for by mythology in the broadest sense. For most people in modern society, science has displaced religion as a creation story. Despite eschewing values and subjectivity in principle, for many it has displaced religion even as a manifesto to live by. Indeed, in spite of itself, science has become something of a new religion, in competition with the older ones. While it may not endorse a morality or ethic, it does embody certain ideals: global brotherhood and cooperation, "ecumenical councils" in the form of world conferences, the common "liturgy" of mathematics, universal institutions and rites of experimental method and peer review, a uniting interest in nature and its authority, and faith in its own precepts and methods. Scientific research is now often conducted by large international teams and scientists typically coauthor across national and cultural boundaries. In many ways the scientific world is a model of social cooperation and decent relationships.

Yet something is conspicuously missing from its account of reality. Scientific objectivity deliberately overlooks the inner life of the subject—his or her embodied and felt *relationship* to the world. But that is exactly what is needed for a mere account to become a narrative that could truly stand in for religion. The border with philosophy would have to become far more permeable. Science would have to embrace goals beyond prediction, control, and the advance of technology, even when these are pursued in the name of disinterested knowledge. By nature, embodiment renders the subject highly interested, and the pathos of the human condition renders the inner life of the subject passionate. Knowledge that cannot encompass these dimensions of living does not represent the whole human being, and cannot serve the greater needs of humanity. Neither can it represent the whole truth of a nature that includes human beings.

Historian of science Gerald Holton identified various cultural themes that underpin the explicit content and concerns of public science, imbuing also its private practice. These are not theorems to be proven, outside time, but rather guiding notions that change with intellectual fashion.[13] The reason such "themata" cannot be proved or disproved is that they serve as axioms, upstream of provable conjectures. In first-order science, they are not objects of debate but background assumptions and rules of the game, which shape the terms of debate. Perhaps the most fundamental of these is the belief that nature can be fully known, formally and exhaustively represented in theory. There are more specific themes, such as the quest for invariants—and propositions based upon them, such as the principle of the constancy of the velocity of light. Themes can make for lively debate, since they often occur in dialectical pairs, but such debate mostly occurs outside science itself. Such pairs include relative versus absolute, contingent versus necessary, reductionism versus holism, realism versus idealism, immanence versus transcendence, indeterminism versus determinism, found versus made.

Modeling and Structure

Scientific analysis attempts to model complex phenomena in simpler, idealized terms. It attempts to express the found in terms of the made, nature in terms of mathematical abstractions or conceptual models. Toward that end, it selects for study phenomena that can be isolated to depend on relatively few factors.

The classical analysis of nature was mathematically framed in terms of differentiable smooth curves, without regard to resources needed to compute such functions except approximately.[14] Instead, aspects of nature were selected that were amenable to such analysis in the first place, sometimes losing sight of the broader goal of "understanding" nature. More recently, other aspects have been recognized that do not easily fit such analysis—for example, fractals and deterministic chaos. These served to widen scientific horizons, for it turns out that chaotic systems are more the rule than the exception in nature. It was systematic disregard of such exceptions in classical thought that led to the preoccupation with determinism and the emphasis on governing laws in the first place.[15] Mechanism was embraced because machines embody systems in which relevant factors are well defined, under human ken and control.[16]

In the eyes of Newton and his contemporaries, the *laws* of nature were mathematical, hierarchical, and external to matter. At the same time, the *details* of the Creation were thought to be contingent rather than logically necessary. This thematic duality was felicitous for the development of science, since it required that one actually observe the details of nature to uncover its order. The laws were to be expressed in equations, especially as functions of time, but the inputs and constants for these equations were to be found by measurement.

Reductionism presents a schematic outline, so to speak, representing the bare bones of nature's *structure*, which is supposed to be what objectively exists. As conceived by the first scientists, it deliberately omits qualities such as color or taste. Yet such qualities turn out to contain information concerning structure. The perception of color, for example, is roughly equivalent to a spectroscopic analysis of wavelength—one aspect of the structure of light; taste is roughly equivalent to analysis of chemical structure. But human sensation does not include, for example, a physiological basis for perceiving polarization. Similarly, scientific cognition is selective, both in its instrumentation and corresponding theory. We can never know that *all* structure of some phenomenon has been identified, let alone all structure in nature. What accumulates in scientific progress is not such knowledge but the empirical evidence on which it is based. New theories must account for the old observations, as well as for the new ones that motivate them. But every new theory can in principle start from scratch in the concepts and the structures it proposes.[17]

Even in their synthetic aspect, laws define *models*, and only indirectly refer to nature.[18] While the model is supposed to map real structure, structural features of the model may be mistakenly imputed to the real phenomenon. There can be perfect isomorphism between *models*, as between mathematical constructs, but never between the model and the inherently ambiguous natural phenomenon it represents. The discrepancy between the model and what it models is considered noise—up to the point when the noise becomes so blatant that a new model must be sought!

Reductionism

Understanding wholes involves understanding the relationships and functioning of parts. In modern times, such analysis refers implicitly to the functional parts of machines—whether physical or conceptual—which are well defined in principle. There is no guarantee that such artifacts correspond to anything in the natural world.

Another type of reduction seeks to understand one category of phenomenon in terms of another of an essentially different sort. Hence the mental is typically reduced to the physical. The implicit assumption is that one category is more fundamental than another and serves as a natural foundation. This assumption, even if by consensus, may be little more than a cultural bias. The downside of this sort of reduction is the same as its advantage: it minimizes the significance of the other category. Is matter more *fundamental* than mind? It is understandable for physicists to think so since their science is devoted to the study of matter. The irony is that physical theory reduces material phenomena to its own mental terms.

Descartes' ambition was to reduce all experienced qualities to extension in space—a program with a long subsequent history in physics, especially reflected in the field concept. This amounts to reducing all sensory information to the terms of the visual sense. Vision was thought particularly reliable partly because light appeared to be transmitted across distances instantaneously. When this turned out to be untrue, the implications in ordinary situations were quantitatively subtle but qualitatively enormous.

Unlike sound, with its longer wavelengths, light allows the information transmitted to be formed as an image, suggesting that the perceived object is what is presented in the optical image. Yet the perception of something cannot be assumed to *resemble* the thing-in-itself of which it is an image. The bias toward vision is not justified by the naïve impression of providing an open window on the world. Rather, it has an *evolutionary* basis, well grounded in the physiology of the anthropoid creature and the physics of its environment.

The ideal of objectivity is that knowledge should be independent of the state of the observer and the path through which it is obtained. Knowledge based on the visual sense, framed in terms of the primary qualities, is assumed to fulfill this expectation better than other sense modalities, which are excluded on the grounds that they lack precision or introduce irrelevant content into knowledge. While the program of dynamics was to reduce all physical quantities to position and its time derivatives, physics actually evolved as a hodgepodge of inconsistent notions. Some (such as "force" and "temperature") derive from other sense modalities, suggesting that the reality of matter could not easily be reduced to mere extension. Descartes felt obliged to exclude force, but the concept of force refers to the real capability of matter to impact the body and of the body to influence other matter. In other words, force implies the reality of matter as it *matters* to the organism. In contrast, visual impressions have no literal impact on the organism, even though they have implications for action and bear information about the interactions of distant things. Part of the problem lay in the very notion of objectivity as knowledge divested of its organic origins. The reality of the material world was supposed to be independent of the embodied human observer, a condition that seemed to be best satisfied by the visual sense. Yet, the real world could literally impact the observer's body and other senses, as experienced in force, temperature, acceleration, inertia, etc. The physics of the world cannot avoid the physical existence and circumstance of the observer, who is *necessarily* embodied.

Reduction to extension—which now usually means some mathematically defined conceptual space—is a recurrent idealist theme in physics. While the reduction is often to a materialist category of thought, the truth is quite the opposite: reductionism itself is deeply idealist—a reduction of the material to mental terms. Extension (a visual quality) is now sometimes in turn reduced to a more fundamental level of mathematical abstraction, just as topology is more abstract than geometry. Space (by any definition) keeps things distinguishable, just as time keeps everything from happening "at once." Yet one should not confuse conceptual space, which is a man-made product of definition, with real physical space, which is something found.

Reductionism is a *theme*, in Holton's sense, imposing constraints on theory. It implies a hierarchy of structure and principles that is not naturally found, but imposed by the theorist (as formerly by the theologian), who decides what is deeper or more fundamental. However, nature may not be hierarchical in that way, but involve processes that are circular and mutually reducible. Unless nature is indeed unfathomable, reduction must bottom out somewhere, bringing to an end the search for ultimate principles. If deeper principles are a function of higher energies, and if this pattern has no natural limit, then the practical limit on available energies could nevertheless impose an end to the search.[19] On the other hand, the ultimate reduction would logically terminate in an entirely self-contained system, offering perfect certainty because it is purely a product of definition. Yet such a system cannot describe *nature*, if nature is real.

A Caveat

Science is no random encounter with the world. Its objects of study are not found raw in the wild, but are carefully prepared in theory and in the laboratory. Data collection is guided by existing theory, and experiments and observations normally look *for* something preconceived. Of course, there have been many accidental discoveries, but these usually occur in the course of pursuing something else. Experiment actively intervenes in natural processes, to re-create them as artificial systems and apparatus. Even in astronomy—hardly a laboratory science—observation is neither passive nor straightforward. Astronomical theory hinges on tenuous chains of inference, based on a few photons arriving from distant space.

Experiment relies on apparatus that is presumed well defined and understood. To the degree that experimental systems may be imperfectly understood, experiment cannot fulfill an unequivocal role as arbiter among theories. Experiments depend on many assumptions and on understanding how the apparatus works within the overall system under study; the interpretation of results is no better than this understanding.[20] The selection of data (against a background considered noise) is also theory dependent, involving judgments open to challenge.[21] Paradoxically, in some historical cases, only the crudeness of the instrumentation could guarantee that patterns simple enough to read as laws could be extracted from the data. Theory would have been seriously hampered if the real complexity involved had been apparent at the outset.[22]

Despite scientific advance, fundamental issues endure, including questions such as: What is the nature of time and space? What are the basic things that exist? Is physical reality continuous or discrete? What is measurement? What is causality? Is there free will? If such questions are not decidable within the scientific framework as currently defined, this is due in part to the nature of the questions themselves; but in part it may be due also to the exclusion of second-order considerations in first-order science. Many contemporary problems in physics have a tortuous history, dating back to speculations of the early Greeks. Such perennial debates often involve logical dilemmas, categories of thought, and the dualism of subject and object. Action-at-a-distance is an example of concepts that are embraced because they work mathematically more than because they make intuitive sense. Glossing over common sense is rendered acceptable by redefining the target system as a deductive system, an idealization *designed* to make sense.

Realism and Idealism

Science is nominally a realist and empirical enterprise, which turns officially to nature for the last word through the testing of theory in experiment. Thus, scientific questions tend to be decidable, while metaphysical ones do not. However, like perception, experiment is no open window on the world but an active intervention involving interpretation. While theory must serve to interpret observation and experiment, it has wide scope to border on metaphysics or fantasy, compromising the realist commitment.

The ontology of science is arguably materialist: all phenomena—including the mental—are supposed reducible ultimately to physical processes. Yet, if science is a cognitive activity, it must include the subject's input, and so it is clearly not driven exclusively by the external world. Its methods and theoretical aspect reflect the creative freedom of the theorist. The ontological materialism of science necessarily rests upon a bedrock of idealism. This is evident in the insistence that physical phenomena can be exhaustively mapped. The faith that reality can be reduced to deductive systems implies that it is *not* viewed as material after all, but as ultimately conceptual.

Realism reflects the natural outward orientation of mind. It serves the organism to know the properties of things that might affect it, and to be able to predict their behavior. The presumption of objectivity, however, says as much about our life as organisms as it does about the world. In truth, it is only conditionally possible to isolate the properties of things—only in regard to idealizations, only relatively, and only in limited contexts. The very meaning of objectivity, like realism, cannot be separated from how it functions for the human organism. Theory always reflects the agency of the theorist as well as the external realities it is supposed to model. Hence, science developed as a synergy of realist and idealist notions, with an ever-shifting balance between them. The realist thread focuses on the role of the external world in supplying empirical data and driving their interpretation. The idealist thread emphasizes the role of the theorist in collecting, selecting, and interpreting data. Neither factor is tenable without the other, for while reality must have inherent properties, these can only be known through creative intervention. In short, realism and idealism are complementary aspects of knowledge, with a dialectical relationship, whose shifting balance gives rise to intellectual fashions.

In keeping with its realism, scientific method is supposed to exclude idiosyncrasies of the observer and artifacts of observation; yet, in realms far from common experience, such effects are unavoidably involved. One is led to wonder, for example, whether quantum uncertainty concerns nature or knowledge. More generally, one may ask whether scientific knowledge is about the *world* or about the observer's *relationship* to the world? Though knowledge of the world must involve the knower, the classical scientific portrait of nature was supposed to be a view of the world, so to speak, when no one is looking. The commitment to realism ostensibly refers to the natural object of study; yet, in practice, it actually concerns agreement among trained observers. This is achieved by minimizing factors that might hinder similar experimental findings by others. But what of factors common to the participation of *all* observers, or prejudices common to a particular generation or cadre of scientists? What of epistemic constraints imposed by nature itself, or arising from the very insistence on quantitative treatment, or from such values as rationality, simplicity, or elegance? The scope and nature of the theorist's participation is excluded from discussion in first-order science, so that such questions are reserved to philosophers.

Today the idealist thread is resurgent in such movements as digital physics and string theory. Perhaps because of competition for research funds, the potential of string theory is ardently contested; yet the disagreement may be less over theoretical detail than over the relative weight of theory in general. It comes down to a disagreement over the very definition of science.[23]

Unbridled idealism considers any impenetrability of nature a temporary inconvenience. Behind this optimism lies the implicit belief that nature lacks a reality of its own that could limit scientific progress. In practice, this assumption has often paid off, both in theory and in technology. Yet, there is no reason, beyond sheer faith, to think that it will in all circumstances. Our broader faith should be in the subject-object relationship and a dialectical balance between idealism and realism.

Notes

1. Taleb, Nassim Nicholas. 2007. *The Black Swan*. New York: Random House.
2. Lindberg, David C. 1992. *The Beginnings of Western Science: The European Scientific Tradition in Philosophical, Religious, and Institutional Context, 600 BC to AD 1450*. Chicago: University of Chicago Press, p. 10.
3. Debus, Allen G. 1978. *Man and Nature in the Renaissance*. Cambridge, UK: Cambridge University Press, p. 2.
4. Holton, Gerald. 2000. *Einstein, History, and Other Passions*. Cambridge, MA: Harvard University Press.
5. For a project to reestablish the genre of natural philosophy, see Unger, Umberto M., and Smolin, Lee. 2015. *The Singular Universe and the Reality of Time: A Proposal in Natural Philosophy*. Cambridge: UK: Cambridge University Press.
6. Jammer, Max. 1957. *Concepts of Force*. Cambridge, MA: Harvard University Press, p. 39.
7. Jammer, *Concepts of Force*, p. 91.
8. Bono, James J. 1995. *The Word of God and the Languages of Man: Interpreting Nature in Early Modern Science: Ficino to Descartes*. Vol. 1. Madison: University of Wisconsin Press, p. 195.
9. Debus, *Man and Nature in the Renaissance*, p. 103.
10. Bono, *The Word of God and the Languages of Man*, p. 240.
11. Debus, *Man and Nature in the Renaissance*, p. 104.
12. For general interest see: Menuge, Angus. 2003. "Interpreting the Book of Nature." *Perspectives on Science and Christian Faith* 55:88–98.
13. Holton, Gerald. 1996. *Thematic Origins of Scientific Thought*. Cambridge, MA: Harvard University Press.
14. Cf. Landauer, Rolf. 1996. "The Physical Nature of Information." *Physics Letters* A 217:188–93.
15. Barrow, John D. 1991. *Theories of Everything: The Quest for Ultimate Explanation*. New York: Fawcett/Balantine, p. 167.

16. Wigner, Eugene. 1960. "The Unreasonable Effectiveness of Mathematics in the Natural Sciences." *Communications on Pure and Applied Mathematics.* 13:1–14.
17. van Fraassen, Bas C. 1999. "Structure: Its Shadow and Substance." PhilSci Archive. http://philsci-archive.pitt.edu/631/1/StructureBvF.pdf.
18. Giere, Ronald N. 2004. "How Models Are Used to Represent Reality." *Philosophy of Science* 71:749.
19. Cf. Horgan, John, 1996. *The End of Science.* New York: Broadway, p. 64.
20. Wick, David. 1995. *The Infamous Boundary: Seven Decades of Heresy in Quantum Physics.* New York: Springer-Verlag, p. 135.
21. Pickering, Andrew. 1984. *Constructing Quarks: A Sociological History of Particle Physics.* Chicago: University of Chicago Press, pp. 5–6.
22. Poincaré, Henri. 1905. *Science and Hypothesis.* New York: Walter Scott, p. 181: "If Tycho had had instruments ten times as precise, we would never have had a Kepler, or a Newton, or Astronomy. It is a misfortune for a science to be born too late, when the means of observation have become too perfect."
23. Dawid, Richard. 2007. "Scientific Realism in the Age of String Theory," p. 8. http://homepage.univie.ac.at/richard.dawid/Eigene%20Texte/13.pdf.

4

Law, Chance, and Necessity

"The law of causality... is a relic of a bygone age, surviving, like the monarchy, only because it is erroneously supposed to do no harm."
—Bertrand Russell[1]

"There is no law except the law that there is no law."
—John Wheeler

Introduction

The ancient dream to know the future demands a foothold outside time and change. If God was privileged to such a perspective, mankind was not—at least not before the Scientific Revolution. The notion of determinism afforded a point of view outside the system under study, from which to regard change without being affected by it. No wonder Laplace had no need for the "God hypothesis," thinking he could access the future through mathematical calculations![2] Yet, one could predict and control such a world only on condition of not being part of it.

The question of determinism versus free will involves the power of one agent to limit that of another. Are one's actions, thoughts, and even intentions autonomous and free? Or are they somehow programmed—by God, by fate, by genes, or by one's political masters? Is one bound, as a cog in a machine, in which the illusion of choice is fixed ahead of time by events beyond one's control? If nature as a whole is a deterministic system, the question cannot be separated from that of one's place within it. As science matured, an image of nature emerged as capable of determining and limiting human experience and behavior. Yet, paradoxically, free will plays an essential role even in deterministic science. For, in order to test theories, the scientist must be free to choose which experiment to perform.[3] Determinism implies a closed system, viewed from outside, and hence the need to stand outside it at least conceptually, if only to supply the input. We are free as well to define the system in the first place.

The very sense of time as a flow into an open future, even from a fixed past, depends on freedom to change the future through acts that are not predetermined. Such freedom might exist either because the world itself is not deterministic or because one lives conveniently outside it. Time is paradoxical for a fatalistic view, which requires some concept of the world in which past and future are mutually entailed.

While unnerving to an agent bound within such a system, the assumption of determinism at least gives the comforting impression that the laws of physics are magical incantations that can be invoked to predict the future. This power seems guaranteed within a deterministic system, where future is mechanically linked to past by known formulae. However, there is no guarantee that the world actually *is* such a system, or—if it is—that the appropriate formulae can be known. Knowledge that enables one to predict the future is based on extrapolating from patterns of the past. While these may be expressed in formulae, no formula has causal power to affect the future or can be guaranteed to describe it accurately.

One could say that *chance* is the scientific version of free will, insofar as both are unpredictable. It can, for example, play the same role in fixing constants or initial conditions that divine will played for the Enlightenment philosophers. Yet determinism and chance are equally ambiguous concepts. What does chance (or accident) mean other than that which does not occur by intention? And what does *determine* mean if not to *ascertain* by an act of free will what is so, or else to *make* it so by an act of free will? Does causality even have a meaning apart from the assessment of conscious agents? Confusion surrounding these concepts leads to the impression that laws of nature "govern," or that mathematical equations somehow carry a power to fix the behavior of matter in advance of actual events. Finally, what does it mean for nature to be "free"—that is, undetermined?

Law

The concept of natural law is ambiguous. It can mean an observed pattern, perhaps for which some mathematical expression has been found.[4] It can also mean some "modal necessity" that *compels* things to follow the pattern and obey the equations. One is a statement of fact, the other of metaphysical fancy.

The very notion of natural law has historical roots in law as *edict*. The two senses of law—as pattern and as decree—have long been intertwined. Certainly, the early scientists made no clear distinction between them and little distinction between causality and will. Newton, for example, was reluctant to attribute gravitation to an inherent property of matter, preferring to see in it the expression of divine will. This was not so far removed from the medieval vision of planets carried around their paths by angels.

While their religious idealism could discount the autonomy of nature, in another sense the early scientists' doctrine of creation indirectly upheld nature's autonomy and favored an experimental science. In order to be all powerful and separate from nature, God must retain the equivalent of artistic license. If the world were strictly a deductive system, as the Greeks had proposed, then God would have had no freedom in specifying its details. A compromise solution to this dilemma was that the Lord designed general laws according to reason but left the details (now referred to as "initial conditions," for example) to his personal whim (now referred to as "chance"). The details of nature remain to be discovered, since divine will was not bound by logical necessity.[5] This stood in contrast to Greek deductionism, which had held that one should be able to deduce natural details strictly from first principles, like theorems of geometry. In that tradition, matter was but an inferior embodiment of the Platonic forms, which could better be known through rational intuition than observation.

Medieval views of natural law were grounded alike in Aristotle and the Bible. Each natural thing was thought to have an innate purpose, as part of the divine plan. Mortals could understand these purposes, and had the free will to respect them or not. This was a *moral* vision of natural law, which—unlike the law of gravitation—could be disobeyed.[6] In contrast, Newton's laws seemed unbreakable, resembling *logical* truths: perfect, infinitely precise, eternal mathematical relationships.[7] While the Deists may have interpreted them as divine edicts, we shall see that such laws attain their proper justification only as limits involving large statistical samples.

A Brief History of Chance

To the ancients, chance represented the will of the gods. Divination attempted to gain information available to the divine mind but normally not to the human mind. It relied on reading events associated with ill-defined objects or processes (e.g., casting bones or stones or examining entrails), which could not be treated mathematically because of their irregular form. The scientific concept of chance begins with what *can* be approached mathematically: some precisely defined regular artifact (for example, the coin with two sides, the die with six equal faces). The modern concept of "prior probability" depends on artificially defined conventions. The roll of dice, for example, is exactly partitioned into equivalent outcomes. Thus, a six-sided die has a prior probability of exactly 1/6 that a given face will be rolled up; this can be confirmed *on average* by rolling the die many times. However, any actual series of coin tosses or dice rolls will not give the precise average predicted. Much less is there a guarantee that nature can be so partitioned, or treated in terms that refer to such idealized situations. Natural situations are not well defined. And theoretical predictions regarding single occurrences must be distinguished from actual statistical patterns observed.

Consulting chance was once a serious theological undertaking, a measure to reveal divine will in time of crisis. The outcome would not have been considered "random" in the modern sense, meaning *without reason or cause.* Quite the contrary, it indicated supernatural intervention.[8] From a modern perspective, such interpretation of chance in intentional terms projects the human desire for order, purpose, and freedom of will.

For early scientists, in their attempts indirectly to read the divine mind, the results of experiment were answers to questions posed to the oracle of nature. The details of nature revealed in such "divination" were first understood as expressions of divine will. Moreover, popular understanding of what we now call chance is not uniformly secular, and remains ambiguous, even superstitious. Gamblers believe in lady luck and some apologists for religion see God's will manifest in what appear to be random events—even, for example, at the quantum level. Such ploys exemplify a long tradition of attempting to assimilate the unpredictable to reason, will, and control.

The modern view is inconsistent insofar as it holds that the universe arose through deterministic processes yet ultimately by lottery. To understand this paradox, it is important to realize that the notion of randomness had been purified of intentionality in classical physics, whereas determinism continues implicitly to involve it. While the biblical creation was a divine act that "determined" the nature of the world in a causal sense, scientific investigation "determines" the nature of the world in an epistemic sense. In either case, an intentional agent is involved, if only to specify the inputs of a deterministic system. The truly random event—such as encountered in the quantum realm—has no specifiable cause. Statistical patterns of events can be observed with precision after the fact, but individual events cannot be predicted.

Lawmakers—such as gods, politicians, and theoretical physicists—are intentional agents, who can serve as first cause to provide the inputs to deterministic processes or reason after the fact what those inputs might have been. But irreducibly random events of nature are by definition thoroughly beyond the grasp of reason and intentionality. As Hume observed, causal necessity is but a kind of shorthand for our natural expectations regarding patterns in the world. Expectation regarding specific individual events is effectively wishful thinking; we are simply dealing with statistics that have no "necessary" basis. In truth, as we shall see, the sole necessity is logical necessity. And even that is not transcendent in any absolute sense. For, in the last resolve logic itself is based on the informal or intuitive sense of what is "logical," which is a highly generalized evolutionary product of accumulated actual experience in the world. Outside the limits of this experience, there is no a priori reason to believe that human logic holds true in all physical circumstances.

This is not to deny, of course, that natural patterns can be effectively modeled. But it *is* to deny that *logical* relations between elements in the model necessarily reflect *causal* relations in nature. All we can be certain of is that correlations exist among *data*; there can always be alternative models to account for them, and there can always be undiscovered correlations. The notion of causal necessity simply transfers, *for psychological reasons*, the logical necessity of the *model* to the physical system concerned.

It follows that *the only truly deterministic systems are deductive systems*. This is because they alone are products of definition, logically closed and fixed, ruled effectively by edict or fiat. Nature and natural systems, in contrast, are open and ambiguous, always eluding definition. It is not nature, then, that is deterministic, but human thought systems projected upon it. Conversely, if it could be proven that nature *is* deterministic, this would imply that the cosmos itself is an intentional construct—a machine or deductive system, a simulation, perhaps an alien engineering project, or a divine creation as originally supposed. In other words, nature itself can be deterministic only if it is not natural!

Governing Power

Nature manifests both random variation and ordered regularity. Patterns in rock formations, for example, are highly varied in detail and yet ordered on a large scale in ways that provide a key to their geological history. While theory can account for large-scale changes, like plate tectonics and the formation of mountain ranges, it does not predict the size, location, and composition of a particular boulder or grain of sand. Theory accounts for the broad strokes but not the details. Patterns expressed by laws may have been generated through historical events (for example, symmetry breaking in the early universe), but the same events are also responsible for random detail. We shall argue that the "immanent reality" of nature resides not in the laws but in the very randomness of the details, precisely because these elude formulation.

While laws of nature properly *express* observed patterns, they are nevertheless sometimes held to *create* those patterns. This notion of physical laws as *governing* confuses the two senses of law. The confusion is plausible enough, given that an algorithm (a computer program, for example) may *compress* empirical data and is also a series of *commands*. Such a program commands the *computer*, however, not nature! Laws and their equations are shorthand for the patterns they describe and expectations based upon them. By imputing causal power to laws or equations, we assure ourselves psychologically of the reliability of the patterns perceived, the validity of our expectations, and our grounds for prediction and action.

Since natural laws codify observed patterns, they follow from the patterns and not the other way around. A natural law is synthetic and inductive; expressed as an equation, however, it becomes a seemingly analytic and deductive formula, with the power to govern the evolution of a natural system in the way that a computer program governs the evolution of the states of a computer. Those are fundamentally different situations, however, unless we assume that nature is literally a machine, deductive system, or computer. The laws of nature are empirical descriptions, which may be conveniently summarized in equations. While a theory may be likened to a computer program, this hardly justifies likening the universe itself to a computer.

Natural laws, like other generalizations, can be *dis*proven, or assumed true, but can never strictly be proven. In contrast, the computer program or equation is a set of rules that are "true" by definition. The computer follows the program because it has been constructed and instructed to do so, whereas nature "follows" laws only in the sense that these are pithy descriptions of what has actually been observed to happen. Mathematical laws describe the behavior of matter but have no power to compel matter to behave in a certain way.

We must conclude that any such power that natural laws might appear to possess derives from transcribing them as *theorems* of a deductive system. Theorems, of course, can be logically proven within the terms of their deductive system. Laws of nature then have an ambiguous status as logical propositions and as empirical patterns. Hume's criticism of causal necessity recognizes that these two very different situations are routinely conflated for *psychological* reasons. The mind attempts to compensate for the ambiguities of knowledge by making precise definitions stand in for the ambiguous realities actually found. So-called metaphysical necessity (or impossibility) is simply logical necessity (or impossibility) projected upon the external world. On the other hand, logic itself is a generalization and abstraction of experience in the real world, translated into a deductive system.

Determinism

In common usage, *determinism* depends ambiguously either on causal antecedents or on possible knowledge of these. In the first sense of the word, some stage in an inanimate process is supposed to have the power to fix the outcome of a later stage; in the second, it is up to some conscious agent to ascertain the facts of a matter. Similarly, "indeterminism" can mean either that we are not in a position to ascertain the facts of the matter, or else that causes do not exist or cannot be meaningfully defined. In the objective sense, determinism means that something is uniquely caused by some definite set of factors. However, identifying causes always involves a selection of some factors or events to the exclusion of others. There is no way to prove that no *other* factors actually enter into play, nor to prove that only a particular sequence of events or set of factors could lead exclusively to a given consequence. On the other hand, there is no way to prove that an event was *un*caused by any factors or prior events whatsoever.

Strictly speaking, however, determinism as causality is a metaphysical idea, apart from which it can only mean logical necessity. Logical necessity is a property of deductive systems and implies nothing about natural reality. It pertains directly to relationships within the deductive system, and only by assumption to relationships within nature. Models may be proposed to rationalize observed correlations, but they are products of definition—made, not found. In contrast, causality is properly an inference from actual observed patterns and measurements. Causality is an empirical issue, while determinism pertains to models. Whatever the quantitative correspondence between the model and patterns of measurements in the real world, there is a qualitative difference between the model and the reality; between them there can be no a priori isomorphism. While exceptions to observed patterns pose problems for theory, from an empirical point of view they merely serve to improve the database.

Determinism in physics suggests that the state of a system at a given time is "fixed" by its prior states. Time is usually presumed, along with a well-defined *system* in which it is a parameter. As we have seen, the word "determine" is ambiguous because it can mean either to *fix* or to *ascertain*. One is an ontological claim, presuming some causal relationship through a physically continuous process between one state and a later one. The other is an epistemological claim, presuming a conscious agent—such as a scientist, historian, criminologist, or judge—who must ascertain the future or past state of a system on the basis of present evidence. In either case, a conscious agent is involved, either to draw conclusions from evidence or else to provide an input for which the deterministic system yields a unique output. That agent could be a researcher trying to find how a physical system responds to different experimental conditions. Or, if the system is conceptual, the agent could be a computer programmer, designer, or theoretician supplying the input to an algorithm. In such cases, it is people, not things, that "determine." In conceptual systems, moreover, logical implication takes the place of temporal succession. Suites of logical operations characterize deductive systems, while suites of physical events characterize physical systems. Time characterizes the real world, while timelessness characterizes deductive systems.

The double-entendre in the concept of determinism is bound up with the uses to which the concept is put and the motivations behind them. The obvious advantage of determinism is to predict the future or retrodict the past. In order to be calculable in the required way, the system must be well defined and follow mechanical rules.[9] In short, it must be a deductive system, which can only correspond approximately to a natural system. Conversely, because the actual correspondence is established statistically, after the fact, one can do no more than assume that it holds in any individual instance or at any given instant. As far as a rational agent is concerned, the found world remains undetermined in either sense of the word.[10] It is only an assumption that its states are fixed in the desired way, and so there is no guarantee of predictability. Nevertheless, there is an obvious incentive to consider nature to be deterministic.

Determinism does not require materialism (nor vice versa), only a unique mapping from one domain to another, regardless of the nature of those domains. As we have seen, causal necessity is but "dimly perceived logical necessity."[11] The precision we associate with determinism is either the precision of definition or of large statistical runs. *All* empirical knowledge is statistical in essence, pertaining to *data* before it makes claims about the world. This is true on any scale; one simply disregards the spread of error in the classical realm in order to imagine a precise and necessary link between cause and effect. What distinguishes the classical realm from the quantum is not the realness of the entities, but the different statistics.

While equations appear to *describe* natural systems, in fact they serve to *define* models that take the place of natural reality. However perfectly mathematics may characterize an idealized model, neither corresponds to reality perfectly. The question remains, Why does the world behave in ways that equations approximate, sometimes very well? Why are observed patterns mathematically expressible at all? An obvious first answer is that we choose patterns to focus on that are easily expressible, with known mathematics. Yet there might be more subtle patterns that go unrecognized, perhaps awaiting future mathematics.

In principle, a deterministic system always evolves the same way from a given starting point. The equations involved in the early studies of dynamics were linear, and well-behaved phenomena were selected for study that could be described by such equations. Hence, the accuracy of classical mechanics is as good as its inputs are precise. However, *perfectly* precise inputs can only be *specified*, never obtained by measurement. Imprecise real-world inputs can lead to exponentially divergent outputs ("chaos") in some systems. Though chaotic systems are considered deterministic—because the *equations* involved are deterministic by definition—those equations are non-linear; this means that small differences of input can result in large differences of output. In contrast, the equations of quantum mechanics (also considered deterministic) are linear, but their output is probability, not individual events. The discreteness of the quantum realm may falsely suggest a special affinity with the discrete states of digital computation, implying that the evolution of physical states can be reduced to computation. Yet the statistical aspect of the quantum realm implies rather that determinism is not the way of nature.

The ability to accurately predict the future of classical systems (such as planetary motions) results from the fact that such real systems coincide, for all practical purposes, with deductive systems. In such massive systems, deviations on a micro scale cancel out to yield a statistically precise averaged macroscopic pattern, which is then codified as a law.

This sort of precision misleadingly suggests determinism rather than an emergent effect of large ensembles of atoms. If the motion of a planet could be made to depend, through amplification, on individual micro events (as in the case of Schrödinger's cat), predicting its path would be impossible.[12] On the other hand, a very large ensemble of such hypothetical unpredictable planets should average out to approach classical expectations.

In ordinary language, *chaos* refers to unpredictability due to lack of order. In science, however, chaos refers to the behavior of systems modeled by non-linear equations. Such a system is deterministic by definition, since it is defined by an equation or computer program (which is how "deterministic chaos" was discovered in the first place). The *computational* processes studied by chaos theory (which presumably apply to many real systems) are non-linear, which means that a later state evolves from an earlier one exponentially rather than in measured proportion. This accelerating change makes such a program very sensitive to initial input. Any slight difference in the input results in enormous difference in the output (the "butterfly effect"). Real systems, such as the weather, *resemble* such non-linear computational processes, and are thus often thought to *be* deterministic, even though they are physical and not mathematical systems. Though the behavior of computer programs can be made to resemble that of physical systems (simulation), the two kinds of things are as different as map and territory. More specifically, they are as different as pseudo-random and true random numbers.

On the other hand, for finite sequences, the distinction between true and pseudo randomness is not absolutely clear.[13] Chaos is interesting precisely because it straddles these categories. It is significant against the backdrop of classical determinism and the modern realization that *most* natural processes cannot adequately be described by linear equations. Perhaps this is because they are not functions of single factors but involve unrecognized multiple causes in complex feedback loops. In any case, this is the way the world really is, in contrast to the simplified and idealized view of classical physics.

If everything is multiply interconnected, then any event is potentially connected to all other events backward and forward in time and at every scale. It will have a bewildering number of causes and effects, with varying immediacy. If all were of equal significance, prediction would be impossible. It's rendered (relatively) possible by ranking significance—for example, by disregarding second-order and higher terms in an equation. A line may be drawn to define a closed system, with denumerable relevant operating factors appropriately defined. Mere physical isolation is not sufficient, however, because even within a limited volume there could be indefinite connectivity. It must also be logically closed, a deductive system.

In summary, the only well-defined systems are deductive systems. The only determinacy is logical implication. The only absolutely certain knowledge is deductive-axiomatic. All physical knowledge is necessarily statistical, involving *some* uncertainty. However perfectly differential equations may describe idealized systems, in the end they correspond only imperfectly to real systems.

The Mathematization of Hazard

Human beings are understandably wary of things they cannot control or understand. In many cases, understanding means modeling a process in a way that renders it predictable. Physical causation may be invoked because it provides a rationale for observed correlations, building confidence that these will reliably continue into the future. For this reason causation is often thought of as a metaphysical power that *makes* things occur. What is actually verifiable, however, is statistical correlation among events that have *already* occurred. Deterministic models are proposed to explain the correlations, but any apparent underlying necessity boils down to the ability of the model to predict the continuing correlation in the future. By positing a cause, or by calling the system deterministic, we merely seek to assure ourselves that the correlation is thus reliable.

Randomness, on the other hand, has long been associated with the darkly unreliable side of nature. In the medieval worldview, chance represented a potentially evil intrusion into the normal (divinely rational) order of things. Though gambling inspired the study of probability, superstition around the notion of chance may help to explain why probability was so late in receiving formal treatment and why it involves persistent confusions.

There are two ways to evaluate likelihood. One is based on past experience, the other on reasoning about idealized situations. One refers to the found, the other to the made. Hence, there are two kinds of probability, which correspond to empirical and theoretical propositions. In the real world, we often need to combine and weigh evidence in a rough and ready way. The sense of the likelihood of an event or proposition may draw intuitively upon relevant past experience, as well as upon some theoretical idealization that is assumed to fit the situation in question. It may also call upon data gathered from others, from observation of natural or social patterns—from questionnaires, for example. Alternatively, it may call upon theoretical reasoning about repeated trials in well-defined ideal situations, such as the toss of coins, the roll of dice, or the operation of roulette wheels.

The idea of the inherent probability of a *single* event is problematic when it refers to real situations. It is meaningful to speak of the inherent probability of a single event only when the context is formally well defined, as in the idealized coin toss. Real events do not usually fall into that category. The notion that a single event is associated with a prior probability is unfounded when it does not refer to a mathematical idealization.[14] Any other estimation of its probability can only refer to the context of a series of recorded past events—that is, a statistic.

An *ideal* coin has an *exact* chance of landing face up, even if it is the only trial of the only coin that ever existed. This is so purely because it is a product of definition. However, this theoretical probability does not predict a single real event or even a sequence of such events, only their relative frequency in indefinitely large samples. An *actual* coin toss is (potentially) a member of such an ensemble of observed actual events. But real coins or dice only approximate ideal definitions. Their imperfections may systematically skew the results—hence the use of loaded dice to cheat.

Estimations of likelihood in everyday situations often cannot be meaningfully quantified.[15] Weighing evidence in courtroom trials, for example, remains an intuitive matter even when it involves formal reasoning. (A jury is bound to a conclusion "beyond reasonable doubt," but not to deliver a verdict of, for example, "76 percent probable guilt.") Logical reasoning about evidence has a long history, invoking diverse interpretations of probability. In modern times, ambiguity concerning probability persists in the interpretation of statistics, which are apocryphally easy to manipulate. A trivial example is the temptation to take life expectancies at face value. By definition, actuarial statistics do not predict an individual's time of death. They are strictly averages of large samples, and cannot foretell your personal fate. Yet your insurance company assesses you individually on the basis of statistics, as though they applied personally to your case. Your doctor may form a diagnosis and recommend a procedure based on precise-sounding figures. Yet your decisions about health care will unavoidably be intuitive and personal.

Probability is a complex issue even in hard science. To rationalize the second law of thermodynamics, for example, it is commonly said that an increase of entropy is not theoretically impossible, merely extremely improbable. To cash in this sort of statement at face value would mean running a virtually infinite series of trials, yet it is confidently made concerning single events. A popular book by a well-known physicist invites us to contemplate the likelihood that an automobile could appear simply by randomly throwing together the atoms that compose it. Regardless of how we explain it, no one would disagree that it is far more sensible to expect that a new car will eventually decay into a pile of rust than that a pile of rust will assemble itself into a new car. But the latter event is nevertheless considered *possible in principle*. For, this author adds, "If you took the atoms apart and threw them together again and again, you'd eventually get a car...."[16]

Apart from the patent nonsense that hypothetical situations often express, such a statement is suspect for another reason involving the subtle assumption that physical reality is reducible to "information." For throwing atoms together in hopes of composing a car is *not* like throwing letters together hoping to compose a message (a topic for information theory). For one thing, the connections between letters that could render them part of a message are symbolic, not physical; they do not inhere in the letters themselves but in human minds. This is in stark contrast to potential connections between atoms. Certain forces, under certain conditions, can indeed lead to self-assembly of atoms. In contrast, it is absurd to suppose that letters of the alphabet have any basis whatever for self-assembling, any preference for each other's company, or anything impelling them to make or preserve bonds. In this mixed metaphor, atoms are to be shaken out like dice or Scrabble tiles, inertly falling where they may, with no basis for recombining other than accidental juxtaposition.

Reordering a set or sequence through random shuffling is an artifice that may not relate to the real world, which does not consist of ideally defined elements. Shuffling a deck of cards enough times *will* eventually produce any given order any number of times, with an average time for this recurrence that depends on the rate of shuffling and the number of cards (Poincaré recurrence time). While each configuration is independent of any other, this is generally not true of physical systems. A container of gas molecules may be a finite collection that is continually "reshuffling," but its possible configurations are not arbitrary or unrelated, as in the case of card shuffling or dice rolling, because the molecules interact with the walls of the container and with each other. It is not valid to argue, for example, that a configuration in which all the molecules end up at one end of the container is inevitable with sufficient time and simply does not occur because it is "highly improbable" at any moment. Unlike cards or dice, the molecules spread out in the container with time, changing their mutual relationships. There may be minor variations in density, but never such a major fluctuation as could (though improbably) occur with a reshuffling. The gas is a physical, not a deductive system.

In the abstract, an automobile too is a deductive system, corresponding to a program for its construction (even atom by atom, granted the requisite nanotechnology). On this view, the program would be a set of instructions—a message—with a certain information content, or complexity. Now, one *can* throw together bits of information (ones and zeros) randomly. In a sufficiently long sequence of such bits, one might indeed eventually get the *program* for the car. However, one would not get *the physical car itself.* For the existence of the car (or any real thing) depends on far more than its information content as defined in the program. However one gets the program—whether randomly or by design—it still must be executed in real materials, which means getting the atoms to relate to each other in just the right ways. Moreover, no part of natural reality can be reduced to a deductive system. While there may be a program for constructing a car, there is no such program to construct a natural thing.[17]

If an automobile is "unlikely" to configure itself accidentally, it would seem all the more unlikely that a whole universe could throw itself together by accident. The probability of a life-bearing cosmos arising "by chance" has been said to be less than 1 part in 10^{123}. However, as we have seen, such estimates of prior probability can only apply to well-defined situations—which the origin of the cosmos is not.

Scientific prediction can be understood either as predicting future events in nature or as predicting the results of observations or experiments. The former can bear real consequences and involve parameters beyond human control. The latter involves essentially well-defined controlled situations. Similarly, probability can be understood in two ways: as a belief about what *will* happen in certain situations (the "Bayesian" interpretation) or as a statement summarizing what has *already* happened in such situations (the "frequentist" interpretation—which may or may not be taken as evidence for what *will* happen). While physicists may have been trained to prefer the frequentist interpretation because it seems to avoid subjective belief,[18] clearly belief is involved either way, and each interpretation is appropriate in different circumstances.

The kinds of well-defined situations to which probability is typically applied usually assume that repeated trials are independent—that is, a later trial is not influenced by the outcome of an earlier one. In other words, there is no feedback. A paradigm example is the fair coin toss. Imagine, however, that every instance where heads comes up in a coin toss somehow increases the chance of heads in future. There is a positive feedback and very soon heads would have a probability approaching 100 percent! (Contrast this with the sensitivity to initial conditions exhibited by "chaos.") Slight tendencies can thereby be parlayed into major patterns, providing a possible means through which nature could self-organize. One could expect interdependence of events to be more typical of complex real situations, involving feedback and multiple causes, than of the simple idealized situations for which probability originated and to which it is usually applied.

Probability is not certainty. There can always be exceptions to the rule, yet one must distinguish between different kinds of exceptions. The "possibility" that a stone might fall upward as a fluke, or as the result of a "statistical fluctuation," must be distinguished from further conditions that can be known to account for such an exception (such as the presence of an overpowering magnetic field). To assign a vanishingly small objective probability to a fluctuation in order to explain its non-occurrence simply mollifies subjective uncertainty about things we can imagine but don't expect. It certainly does not mean that gravity operates only a very large percentage of the time, outside of which things are able to fly up.

Ensembles and Single Events

In the quantum realm, probability properly refers to ensembles of experimental events. The probability of a given event can only refer to frequency within such runs. Though individual events are thus "random," there is an overall order. While classical statistics yields the characteristic Gaussian (sine-shaped) curve, quantum statistics yields a modified pattern, characterized by superposed curves (interference).

Even in the classical realm, the patterns we know as laws—and prediction of individual events—properly refer to members of an ensemble (of macroscopic measurements, for example). In both domains, therefore, physical law refers to reproducible events, hence to ensembles rather than individual occurrences.[19] Yet classical randomness (as in the toss of a coin) is held to be of a different sort than quantum randomness (as in spontaneous decay of a radioactive atom). Classical randomness is pseudo-randomness, insofar as the outcome is held to be determined by an equation of motion—governing an idealized coin, for example. The outcome is supposed to be unpredictable in practice only because the real initial state cannot be known precisely enough. Yet it is no coincidence that the coin is an artifact, an object made to conform as nearly as possible to an ideal definition.

Now, if determinacy does not apply to nature, neither does indeterminacy. Just because no cause *has* been found for an event (ergo, no basis for predicting it) hardly means that no such cause might be found in future. However, since it is impossible to *prove* that no cause whatever can exist for it, it is impossible to prove that an event is intrinsically without cause—that is, "random." On the other hand, it is equally impossible to prove a given cause for an event, which could always involve coincidence or additional factors. Hence, it can neither be proven that a specific event is caused or uncaused. It might seem that one is on safer ground to consider only sequences or ensembles of events, which can be tested for patterns or order. But a similar difficulty occurs then too since failure to find a pattern does not prove its nonexistence.[20] Passing every known test for randomness would not strictly prove that no algorithm exists to generate it. Randomness has no rigorous meaning apart from such tests. While it would be convenient to point (as I do) to randomness as a signature property of nature's autonomy, the situation is less than straightforward, if only because randomness cannot be proven. Nevertheless, Gödel's theorem suggests that unpredictable events must occur, and there must exist natural facts that cannot be accounted for by any particular theory.

Order and Complexity

In his *Discours de la Métaphysique,* Leibniz had proposed a thought experiment to explore what sort of thing should be considered a law of nature. Splatter some dots of ink randomly on a sheet of paper. Note that they can always be connected by some line, and that such a line can always be described by some equation. However, only if this is a *simple* equation, Leibniz argues, can it be called a law of nature.[21] Mathematician Gregory Chaitin developed the idea that the degree of complexity of the equation connecting the data points can be measured by considering it to be a program that generates those points. Only if the program (expressed as a string of bits) is significantly shorter than the data set itself (also expressed as a string of bits) is it a "law." The best theory or law is the smallest program that generates the data.

Intuitively, one thinks of complexity in terms of things that are complicated and ordered, for example, a modern car. Yet we have seen that machines and other artifacts are actually simple; their order is produced by defined procedures. Somewhat paradoxically, scientific definitions of complexity are closely related to *dis*order. Chaitin's definition, for instance, measures the complexity of a system by the minimum length of computer program that can describe it. By this definition, a randomly generated text is more complex than one deliberately composed![22]

Many scientific definitions of complexity involve additional concepts, such as order, entropy, or information, which are no easier to appropriately define. Complexity is sometimes associated with the entropy of a system (its disorder) and sometimes with regularity or structure (its order); sometimes with novelty (information) and sometimes with noise (lack of information). Just as it makes no sense to speak of the probability of an individual digit apart from how it is generated, so it makes no sense to judge the complexity of an isolated structure or process apart from the processes that gave rise to it. In other words, the appearance of order depends on context and history. A pile of books on the floor may appear disordered, compared to neatly shelved books in alphabetic order. However, the appearance is relative to intentions behind the books' placements. If they had been carefully placed on the floor according to their relevance in a research project, for example, their order would be greater or more significant than if they had merely been alphabetized on the shelf.

Simple natural forms tend to be robust insofar as they are favored across a variety of conditions, while more complex forms may be more sensitive to disturbances or to initial conditions.[23] Yet systems sensitive to initial conditions seem to be far more prevalent in nature than formerly thought. Their neglect before the invention of the digital computer reflects a bias toward idealizations that can be treated mathematically with paper and pencil: stable linear processes, closed reversible systems, and so forth.[24] Yet deterministic chaos is not the end of the story. For, other systems are robust *because* of their extreme complexity, the paradigmatic example being living things.

Natural complexity seems to build up incrementally over time.[25] The complex systems we are most familiar with are living organisms, so it is natural to think of the evolution of complexity in biological terms: Simple forms evolve, through mutations and selection, into more complex ones. Yet how such processes might apply in the nonliving world remains to be elaborated. There is more to the question of evolution, in nonliving matter as in living matter, than random mutation and natural selection, which are passive mechanical processes. In particular, principles of active self-organization must be brought to bear, such as in a generalized theory of autopoiesis.[26] This stands in contrast to conventional reductionist ideas. Like "chaos," it may turn out that complexity—not simplicity—is what we shall find about us "fundamentally" in nature.

Notes

1. Russell, Bertrand. 1913. "On the Notion of Cause." *Proceedings of the Aristotelian Society* 13: 1–26.
2. Pierre-Simon Laplace (1749–1827), French mathematician and astronomer. His famous reply to Napoleon when asked why Laplace had not mentioned God in his writings on astronomy: "*Je n'avais pas besoin de cette hypothèse-là*" (I had no need of that hypothesis).
3. Gisin, Nicolas. 2010. "Is Realism Compatible with True Randomness?" http://arxiv.org/pdf/1012.2536.pdf.
4. The general idea of the mathematical *function* was a prerequisite for the notion of physical law, as developed in the seventeenth century. See Whitehead, Alfred North. 1925. "Mathematics as an Element in the History of Thought." In *The World of Mathematics*, vol. 1, edited by James R. Newman. New York: Simon & Schuster 1956, p. 411.
5. The position here, of course, is that laws are merely generalizations of such details in the first place.
6. Thompson, David L. 1991. "Concepts of Nature: Are Environmentalists Confused?" www.ucs.mun.ca/~davidt/Nature.html.
7. Davies, Paul. 2011. "On the Multiverse." FQXI Setting Time Aright Conference. www.youtube.com/watch?v=ulkulX8O5lI.
8. Barrow, John D. 1991. *Theories of Everything: The Quest for Ultimate Explanation*. New York: Fawcett/Balantine, p. 173. We moderns retain the idea of natural disasters as "acts of God."
9. Earman, John. 1986. *A Primer on Determinism*. Dordrecht, NL: D. Reidel, p. 111.
10. This does not mean that it manifests some fundamental property of "indeterminacy," which is another metaphysical notion.
11. Earman, *A Primer on Determinism*, p. 98.
12. Ellis, George F. R. "Physics in the Real Universe: Time and Spacetime." Sec. 3. arxiv.org/pdf/gr-qc/0605049v5.pdf.
13. Chaitin, G. J. 1999. *The Unknowable*, New York: Springer-Verlag.
14. It is merely by convention that, *in the absence of further information*, an event has a "50 percent chance" of happening. And one cannot retroactively assign a prior probability to a past event, for the probability of any event that has actually occurred is 100 percent!

15. Franklin, James. 2001. *The Science of Conjecture: Evidence and Probability before Pascal.* Baltimore: John Hopkins University Press, p. xi.
16. Susskind, Leonard. 2008. *The Black Hole War.* New York: Little, Brown, pp. 128–29.
17. The genetic code is a set of instructions to the natural environment, not to the microbiologist or genetic engineer.
18. Baeyer, Hans Christian. 2013. "Quantum Weirdness? It's All in Your Mind." *Scientific American* 308:46–51.
19. Cf. Jauch, J. M. 1973. *Are Quanta Real?* Bloomington: Indiana University Press, p. 47.
20. On the other hand, though an algorithm to fit a finite sequence might be found, it still could be coincidentally part of a longer sequence for which no algorithm can be found.
21. Chaitin, Gregory. 2008. "Leibniz, Complexity and Incompleteness." www.academia.edu/5802738/Leibniz_complexity_and_incompleteness_Rome_.
22. Horgan, John. 1995. "From Complexity to Perplexity." *Scientific American* 272:106.
23. Laughlin, Robert B. 2005. *A Different Universe: Reinventing Physics from the Bottom Down.* New York: Basic Books, p. 130.
24. Barrow, *Theories of Everything*, p. 167.
25. Smolin, Lee. 2013. *Time Reborn.* Toronto: Alfred A. Knopf Canada, p. 194.
26. Following Maturana and Varela (*The Autopoiesis of Cognition.* Dordrecht, NL: Reidel, 1973), autpoiesis means "self-making," which includes not only self-reproduction and self-maintenance but also "self-definition."

5

Mathematical and Physical Reality

"Mathematics is a great tool, but the ultimate governing language of science is language."
—Lee Smolin[1]

"As far as the laws of mathematics refer to reality they are not certain; and as far as they are certain, they do not refer to reality."
—Albert Einstein

The Reasonable Effectiveness of Reason

Physics relates natural realities to mathematical idealizations. But what exactly is the basis of that relationship? In particular, why is mathematics so successful at predicting the behavior of matter in unfamiliar situations? Until recent times, the question did not engage many physicists, who were busy simply taking advantage of that success. This omission probably owes something to the early alliance of science and religion, which took for granted a divinely ordained correspondence between nature and reason. Yet the question could shake the very foundation of physics. Some answers could undermine the hope of formalization that is so dear to the theorist; others could threaten the realism that is the very basis of experimental science.

Leibniz took note of the apparent correspondence between the mental and physical realms—in particular, between mathematics and the physical sciences—which he took to be an act of God, a divinely "pre-established harmony." In a more secular age, physicist Eugene Wigner would famously call this harmony "the unreasonable effectiveness of mathematics."[2] In the era of computers, the question has regained interest among physicists, mathematicians, and philosophers, not to mention science fiction writers. Why then *is* mathematics so effective at describing nature? Opinions range from viewing the universe as an inherently mathematical structure to viewing mathematics as an inherently biological phenomenon—and various shades between.

Let us begin by acknowledging that (whatever else it is) mathematics, like physics, is a human undertaking—in contrast to nature, which is not. The position taken here is that mathematics is a form of cognition; at first glance, its effectiveness should be no more surprising than that of ordinary cognition. It effectively describes the real world at least in part because it abstracts the world's most general properties in the first place.[3] While mathematics is logically self-contained *as a conceptual system*, its development *as a human creation* is informed by categories and relationships gleaned from physical reality. Just as mathematics is not obliged to copy nature, so nature is not obliged to follow the mathematical development it has inspired; but neither should we be surprised when she does. After all, the properties of integers reflect the integrity, permanence, grouping characteristics, etc., of real things. The operations of arithmetic and the properties of sets generalize our basic expectations concerning objects and relationships in the world. The salient question is why mathematical intuition should continue to apply to realms that are far beyond our ancestral and even present everyday experience.

Let us grant that both logic and mathematics do reflect and generalize aspects of natural reality.[4] Their abstractions are then projected back upon nature as underlying its features and are also reflected in the ways experiments are constructed. The expectation that nature will behave mathematically is then something of a tautology. Is it a miracle that equations, which idealize the characteristics of objects and space, happen to describe experimental setups that are also idealized? To be described mathematically at all a natural system must first be thus redefined. The question is then: Why do these idealized constructions correspond so closely to natural realities? One answer is that there is a selection effect involved, so that reality appears intrinsically mathematical because we focus on those aspects of it that can be most easily treated mathematically, or which correspond to the behavior of experiments. Idealizations are chosen that *do* correspond closely enough to reality, or which are assumed to by convention.

If we marvel at how effectively simple mathematical expressions portray the real world, we should bear in mind that far more complicated expressions for the same relationships can be found. We latch on to the simplest expressions, and are pleased by their effectiveness because such equations are succinct and solvable. We choose to look at simplified aspects of nature because these are what we can deal with mathematically and technologically. A further benefit is that conceptual machines can lead to the construction of real machines, such as computers and radio telescopes. However, the apparent mystery (that computation mirrors the behavior of real things) puts the cart before the horse, for computation is simply the latest mechanist metaphor by which to understand the appearances we call nature.

While we marvel at the effectiveness of mathematics to model nature, perhaps the true marvel is the general ability of thought to model the external world at all: the preestablished harmony between world and brain. Einstein considered the capacity of reason and mathematics to grasp reality to be miraculous since (following Hume) he saw no necessary connection between experience and reality. Historically, what was missing in his generation's understanding of that relationship is what is today called evolutionary epistemology: connections between experience and world are established through natural selection. If such connections are historical, and not logical in some a priori sense, then logic itself must be an evolutionary product. An evolutionary general theory of intelligence might help account for the astonishing effectiveness of mathematics, which would then be one development of a broader capacity to model, abstract, and generalize.[5]

The correspondence between mathematics and physical reality is satisfying for psychological reasons as well as for the competence it affords. We are pleased to think that physical reality is rational and can generally be redefined in human terms. We are put at ease in believing the world corresponds to our definitions. Successful mathematical treatment of nature, reflected in technology, richly confirms that faith.

Like the ambiguity in consciousness itself, the truths of mathematics are of a dual nature, referring both to the world and to inventions of the mind. Like all forms of cognition, mathematics reflects both the object and the subject, and represents external reality by translating it into its own terms. While it is amazing that nature appears to follow its formalisms into unfamiliar territory, is mathematics inherently better equipped than other forms of cognition for this purpose? After all, the power to generalize and extrapolate from the familiar is universal, allowing the extension of knowledge through metaphor and analogy. While mathematics may be viewed as a sort of language, ordinary language itself is essentially metaphorical. However, a first answer to the above question is clearly yes. By virtue of its very generality mathematics *is* uniquely suited to transcend circumstance and scale. Yet this answer must be qualified, for if nature appears to follow the indications of established mathematical and logical principles, it does so to an unforeseeable extent. In some cases nature *leads* the way to the creation of new mathematics and even logics. Just as novel experience, heretofore unexpressed, sometimes gives rise to new language, newly discovered aspects of physical reality—which do not yet correspond to known math—sometimes suggest or require future developments in mathematics. The universality of mathematics is certainly plausible in view of the apparent isotropy of the known universe and the self-similarity of nature at diverse scales. To take these for granted, however, would beg the very question at hand: to what extent, and under what conditions, can general principles be extended into unfamiliar domains?

Metaphor is always selective, revealing certain aspects and concealing others, making it risky to confuse the metaphor with what it represents. Assumptions based on experience on the human scale may or may not transfer to the microscopic world, to the cosmos at the largest scales, or to the realm of the very complex. It is merely a working assumption that the same mathematics should apply in every new situation.

The Hazards of Mathematization

Models are deductive systems, usually defined by equations, which are used to represent selected aspects of the real world. Strictly speaking, they represent *records of specific interactions* between observer and world.[6] The equations and the models are isomorphic to each other, since they express the same idealization. However, neither can be assumed to be isomorphic to the real process or system modeled.[7] Even if a deductive system is complete in the mathematical sense, it cannot represent *reality* in a strict one-to-one correspondence. In this respect, scientific modeling is no different from other cognitive processes that are highly selective, serve biological needs, and are limited by the intentions behind them and the material processes that support them. The very idea of mathematical prediction, for instance, presupposes an agent with reasons to predict the future and values that inform those reasons.[8]

The correspondence of mathematics with physical reality parallels the correspondence of our perceptual models with the external world. In the case of perception, such modeling serves the creature and is endorsed by its evolutionary success. However, mathematical modeling, though highly useful, has not been around long enough to prove its long-term evolutionary value. While our perceptual models are so ingrained that we take them for the world itself, taking mathematics for the world itself does not enjoy the same warrant. It suggests, nevertheless, that physical reality is reducible to math. Or it suggests that the domain of math is a higher version of reality, with its own sort of substantial being, which exists prior to and independent of physical reality. Accordingly, one explanation of the miraculous correspondence between mathematics and nature is that they are simply one and the same: the universe is "really" nothing other than a mathematical structure, a Platonic form. A variant of this idea is that only a subset of mathematics is endowed with the mysterious "property" of physical existence. When mathematics and thought are thus assumed to be primary, they can appear to have causal powers. The world is then a byproduct of first principles, hence, the compatibility of Platonism with early Christian dogma, which expected a fit between reason and the world as a creation of the divine Mind.

The mathematization of nature can be misleading in a third way we have already mentioned: it restricts the operative factors involved to the few that can be treated mathematically. While this is useful "for all practical purposes," it creates the misleading impression that natural reality involves *only* these defined factors. While it may be true that science as we know it could not have developed without such restrictions, this does not make it a true picture of the world, nor should it constrain the science of the future. We do not know what purposes in future may be deemed "practical."

The fact that the *mathematics* seemed to work precisely in classical mechanics was taken to mean that *causality* worked perfectly behind the scenes. This amounted to the realist assumption that physical variables must *have* precise values, even when they could only be *measured* imprecisely.[9] But physical variables and their equations are idealized constructs. While they are well defined by fiat, their presumed ability to predict accurately in a given instance is an act of faith.

Natural laws are often expressed as mathematical functions of time. Can the equations themselves change over time? Changing natural laws can, of course, be incorporated as time derivatives of the original equations. But since a deductive system has the connotation of being timeless and self-contained, the psychological implication is that changing laws are not to be expected any more than unforeseen variables are to be expected in a theory deemed complete. In other words, the sheer logical closure of deductive systems may falsely lead to the expectation of a final theory.

As Spinoza recognized, mathematics is a less than reliable guide to nature when it imposes its own consistency, harmony, and elegance.[10] Symmetry, for example, is a *mathematical* property, with a strict definition, that also describes some physical things and processes. While physical symmetry is ever only approximate, mathematical symmetry is abstract and precise. One is empirical, the other a matter of definition. The success of mathematical symmetry arguments at predicting discoveries in high-energy physics suggests that symmetry (of laws, at least) is an objective property of nature. Yet this does not mean that nature *must* be profoundly symmetrical at deep levels simply because the mathematical theory of the day involves symmetries. One should also bear in mind that "new" particles (as predicted through symmetry arguments) are not naturally occurring entities but products of experiments.

"Symmetry breaking" describes the departure of reality from the ideal. That is, when symmetry is considered the ground state, an asymmetry is sought to account for any departure from it. (The concept may be compared to Aristotle's notion of a natural way things are, which can be disturbed by an external efficient cause.) In modern physics, such a view is prejudiced in favor of artificially defined symmetry; it reifies asymmetry as a causal agent, whereas symmetry and asymmetry are but ideas imposed by the theorist upon natural situations, which are generally neither static nor symmetric.

In general, mathematization creates a parallel reality with its own rules, as we will see in the treatment of various conceptual spaces. Aside from these few examples from physics, there are far broader ways in which mathematics makes over the world. While it is embraced as the language of science, it may be misguided and even harmful as the language of economics, for example, leading to an undesirable standardization of human values and relationships. Money quantifies, standardizes, and objectifies "value," which nevertheless remains inherently subjective. The quantification of value, and of the intentions that give value to human labor, detract from appreciation of quality—of things made and services rendered. It leads also to depersonalization of human effort and experience as just another commodity, another resource to exploit. The fact that economics deals with complex interactions of many agents, involving convoluted feedback, means that a linear approach cannot work even in principle.

Platonism

Galileo famously called mathematics the language in which nature is "written." But atoms are not literal numbers and nature is not literally a text. Mathematics is rather the language of science—or at least its grammar. There is no more guarantee that science and mathematics can capture all aspects of natural reality than there is that ordinary language can capture all of human experience. And mathematics resembles a game as well as a language, in which possible moves are defined by convention. While we expect nature to abide by its own rules, in fact we know only the rules we have devised, which may not fit nature within some situations or in all ways.

The relationship between human conception and reality resembles that between the digital and the analog. On the macroscopic scale at least, nature appears "dense" in the way the real number continuum is: as an analog domain. Products of definition, on the other hand, are "sparse" in the way that the natural numbers are sparse compared to the real continuum. Yet while the real numbers cannot be mapped onto the natural numbers, they are a logical development of possibilities inhering in them. While the unreasonable effectiveness of mathematics could be interpreted to imply that mathematics preexists nature in some sense, it could equally be interpreted as evidence that nature has preceded mathematicians in the expansion of "logical possibility"—which is itself a generalization of possibility experienced in the natural world.

The belief that mathematics exists apart from mathematicians, and so is something they come upon as found, stems from the conviction that logical possibility is not a matter of empirical experience, but something transcendent, eternal, objective, fixed, and independent of human thought and perception. A Platonic interpretation of mathematics regards the totality of mathematical possibility to be a finished and timeless structure. While Gödel clearly believed in such a thing, ironically he showed that any constructed part of mathematics (of a certain complexity) cannot map the whole. Because imagination is self-transcending, the *whole* of math cannot be contained in formal definitions. Though a given mathematical construction may be "timeless" as a product of definition, someone makes it at a particular time and place. The corpus of mathematics has a history and is a work in progress, subject to indefinite expansion and refinement. Mathematics is a matter of formal definition and of logic, neither of which is fixed in advance or for all time.

Nature cannot be contained in formal definitions because one cannot anticipate what may yet be found. By definition, any formalization of a natural phenomenon is isomorphic to *some* mathematical structure. But there is little justification, except psychologically, to identify the whole of nature with the whole of mathematics, when it is not possible to exhaustively specify either.

Real and Conceptual Space(s)

The general concept of dimensionality in modern physics conflates physical with mathematical space. Yet physical space is a phenomenal fact, whereas mathematical spaces are conceptual artifacts. Their very advantage is that—unlike real space—they are malleable products of definition.

The mathematical notion of dimensionality in physics is grounded in the Cartesian coordinate system, itself based on solid geometry. The conventional three dimensions of geometry are abstracted and generalized to include any number of causal factors, characterized as the "dimensions" of some conceptual system. The joint operation of such factors defines a "configuration space," which is often used in conjunction with a prepared experimental setup. The question of what real space "is," and how it might vary or evolve, is difficult to separate from such conceptual tools.

To treat space as an abstraction, with an arbitrary number of dimensions, allows a number of interesting but peculiar questions to be raised. For example: How does the world happen to have *three* spatial dimensions when any number is theoretically possible? Do the laws of physics entail a particular number of dimensions? Does space have the same dimensionality everywhere and at all times? Is the (three-) dimensionality of space an illusion? However, the fact that such questions can be posed does not mean they make sense or have meaningful answers.

The idea that the tridimensionality of physical space is something to *explain*, or that it could be otherwise, confuses real with conceptual space. While conceptual space can have any number of dimensions, it is only by *convention* that real space has any "dimensions" at all. The notion that real space has three dimensions comes of treating it geometrically, since only three lines can be mutually perpendicular. However, it is merely a convention that higher dimensions in abstract spaces are "orthogonal," on the metaphor of Euclidean geometry. In abstract spaces, it simply means that the relevant variables are defined to be independent. However, it can never be known with certainty that two such variables are in truth completely independent.

The question of why we live in a world of three dimensions treats dimensionality as though it were a *physical property*, whereas it is a matter of convention and definition. Some equations of physics can be made to support a number that differs from three, or to retain the same form for a varying number of dimensions.[11] This is supposed to demonstrate that dimensionality is an unknown physical quantity to be determined by experiment or, ideally, to be "predicted" by theory. However, properties of *equations*, defined for a conceptual space, do not necessarily imply anything about physical space.

The "block universe" (Minkowski's four-dimensional space-time continuum) is a convention that unifies abstract concepts of space and time, which—in accord with special relativity—no longer have independent significance. In effect, it treats time as a fourth spatial dimension. It is a static or "frozen" view of change, perhaps psychologically motivated by the desire to transcend change in a godlike perspective. The *mathematical* motivation of the space-time continuum arises from the fact that certain physical concepts can be treated more conveniently than by dealing with space and time separately. To consider this convention as *real*, however, is a metaphysical assertion.

The mathematical space-time continuum employs a space-time interval rather than a space interval *through* time. This interval has the speed of light built into it, and so incorporates the effect of rapid motion and is invariant among observers in different reference frames. However, the space-time interval can only be *calculated*, using the Pythagorean theorem, on the basis of space and time intervals separately measured, using rulers and clocks. Except for light signals, no measuring tool exists that is a hybrid of a ruler and a clock! But to *define* light as the measure of this interval is circular reasoning, since the speed of light itself is defined in ordinary units of space and time.

As Heraclitus observed, it is not possible to step into the same river twice, unless, of course, it is an idealized, static, mathematically defined river! Even the notion of time as a flowing current is already an abstraction, insofar as past, present, and future are conceived on an equal footing as points along a "continuum." In phenomenological terms, however, the present is one's waking experience, the past is memory, and the future is imagination. The fact that we can anticipate future events at all seems to reflect a real continuity in the world but does not imply a flow of "time," which is not a thing but a nickname for that experienced continuity. For the same reason, we intuitively abstract "space" as a container, backdrop, or stage for experienced events. This too is but a nickname for the experienced separation of things. Events may succeed each other either through motion or through intrinsic change. Yet in phenomenological terms, there are simply changes in a sensory field, which are interpreted in various ways—for example, to establish "here" and "there," "now" and "then," "objects" and their "motions," and transformations through "time."

The Physical Context of Knowledge

The world is intelligible because of a synergy of its real properties with successful strategies employed to model and understand it. The extreme speed of light, for example, had made it possible to ignore the very small relativistic effects occurring at the speeds that could be studied during the first two centuries of classical physics. Similarly, the very small size of atoms, electrons, and energies of visible light made it possible to ignore quantum effects among classical objects. While these circumstances made the development of physics possible in the first place, they also required eventual revisions and posed limits on physical knowledge. On the other hand, if it had been the case that light traveled with infinite speed, as was early supposed, life would not have been shielded from the simultaneous arrival of all the radiation and other distant influences that might occur at a given moment. Similarly, but for being quantized, matter would not be stable; chemistry would not be possible, and neither would chemists!

If life has depended on relative isolation and separation of effects on different scales, so has science.[12] Without such localism, the notion of the isolated system would not have been feasible and physical laws could not have taken a simple and practically computable form. These were also facilitated by the apparent continuity of the macroscopic world, for which differential equations were effective.[13]

The huge difference in size and energy between macroscopic objects and photons allows one to neglect the physical effects of observation in many circumstances. Because of this disparity of scale, one can postulate the existence of real objects independent of observers (and of observers independent of objects). Without this effect of scale—which might be but an incidental fact of our universe—there could be no clear distinction even between subject and object. In that sense, philosophy is at the mercy of physics: such categories cannot be taken as a priori if they depend on contingent physical facts.

So-called anthropic reasoning[14] is an example of a wider type of consideration that is hardly restricted to human observers: those creatures exist whose existence is permitted by the specific nature of the world and their behavior within it. The fact that life (with scientific observers) presently exists implies, trivially, that the universe is such as to have allowed its evolutionary history. Without concluding that the universe aims toward life, one could plausibly go further to speculate that self-organization in the universe at large was a necessary backdrop to the self-organization of life.

Discreteness

Nature obviously has discrete aspects at various levels. Macroscopic objects are resolvable into smaller objects, and so on. One can *postulate*—but hardly prove—a fundamental bottom level. Furthermore, discrete *states* can often be regarded as effects within some continuous field. In any case, *physical* discreteness is categorically different than *mathematical* digitation. While the quantum realm *seems* to correspond to precisely defined theoretical constructs, there could be finer-detailed levels of description (i.e., so-called hidden variables).[15]

The discrete states of a physical device, such as a digital computer, are thought to depend ultimately on quantum discreteness.[16] Yet discreteness is literally a matter of definition. Classical machines can be in discrete states because their parts are at least relatively well defined and stable. (That is, they correspond to their design well enough for their purpose, so that computers, for example, can transmit discrete states reliably.) At a finer scale, however, the parts of machines are fuzzy; when they do not correspond perfectly to their ideal definitions, transmission of discrete states may be unreliable. While macroscopic physical systems correspond only approximately to their theoretical ideals, at the microscopic scale one presumes that physical entities are *identical* to their theoretical counterparts, so that the discreteness of real particles and states corresponds *by definition* to their discreteness in theory. Yet this is merely an article of faith.

Quantum entities have properties, such as electric charge and spin, which define their kind and don't depend on their momentary state. (They also have other properties, such as position and momentum, which do).[17] In the quantum realm, there are difficulties of principle involved in finding all the objects present of a given kind (the extension of the set).[18] And it is merely an assumption that elementary particles have no *other* properties than those that currently define them (the intension of the set).

Particles are indistinguishable when we cannot point to them individually, or to individuating properties, as we can in the case of planets, chairs, or other macroscopic things. However, such restrictions do not imply that they are mere products of definition—that is, essentially mathematical objects—any more than planets are. Even the identity of celestial objects often depends on the sort of detection events encountered with elementary particles, yet we do not conclude that such objects are less than fully real.

Some light is shed on the conundrums of individual identity by likening quanta to dollars in a bank account.[19] A detection event (observation) is like the act of withdrawing funds from a bank account; it occurs at a given time and place and bears a distinct record (withdrawal slip). However, though the money involved is a quantized value, it does not consist of *specific* dollar bills or coins. Similarly, though energy is quantized, it makes no sense to speak of identifiable individual quanta of energy.

What is "Information"?

The intuitive concept of information ambiguously reflects both reality and the purposes and nature of communicating agents. It is misleading to objectify information simply on the grounds that it reflects real structure, without considering also how it reflects the agents perceiving and conceiving it.

In the biological sciences, information is a useful concept in the explanation of cognitive behavior. Yet it is even less appropriate there than in physical sciences to objectify it in a way that divorces it from meanings and the uses. In neither the physical nor the biological sciences can information stand on its own as a causal substrate, or serve as an ultimate ontological basis for reality. It cannot be quantified in an absolute sense, but depends on how the situation is partitioned; nor should it be treated as a causal agent in circumstances where *we*, in fact, are the causal agents. On the other hand, the very ambiguity of the concept helps straddle the gap between mind and matter/energy. Since meanings must be carried on physical signals, information is both mental and physical.

Some thinkers see information "encoded" everywhere in nature, without asking who encodes it and for what purpose. Allusion is made to information being stored, encoded, or processed in natural systems—implying a form of computation, but without discussing the agents doing the computing. (This is sometimes a mere trope, when the author is actually speaking of causal continuity.) Organisms do explicitly need to model their environment, encoding it in some representation and storing information in memory. But it is trivial to assert that physical reality at large "computes" itself, and it is metaphysically unjustified to assert that nature has any need to represent itself. Organisms encode information for their purposes, and physicists encode it for *their* purposes.

Only products of definition have a definite information content. A model, formalism, or theory contains definite information to the extent it is well defined. However, no *real* system is inherently well defined. It cannot, therefore, have a definite information content—even if there is a fundamental bottom to natural complexity. Yet this has become a basic assumption of contemporary physics, as expressed in the notion of a quantitative information bound.

How many bits of information are there in the universe? Any answer to this question depends on how information is defined and measured. Concerning physical reality, information must correspond to physical structure. Is there, then, a limit to how much structural detail can actually exist in the universe? Or is the limit rather on what we can know of this structure? It is reasoning in circles to decide that the universe contains a finite amount of information because there must be an absolute bottom to its structural detail. (And vice versa: it is circular reasoning to decide that there is a bottom to its detail because it must have a finite information content.) Alternative assumptions include the possibility that natural systems do *not* have definite information content at all, that information bounds are theoretical artifacts useful in some contexts and perhaps not in others, and that there is no intrinsic bottom to the complexity of nature. To pursue such alternatives as working assumptions could lead to a different view of the constitution and history of matter.

What, then, *is* information? Clearly it *informs* some agent about some difference, for some purpose, in some context. Because it is *for* an agent, it cannot be a matter only of "objective" description. Its treatment in physics and engineering should not obscure the fact that it refers both to subject and object. Yet it is usually held that information as defined quantitatively must be distinguished from meaning.

The Shannon information content [20] of a message belongs in a different domain of description from its meaning, reflecting the difference between syntax and semantics. It is supposed to be free from *particular* meanings and *particular* agents. This, however, does not imply it can be free from meaning and agency, *tout court*. Mere abstraction does not free information from communicating agents who can slice up the world in diverse ways any more than it frees mathematics from mathematicians.

While information refers to real structure, to some extent structure is in the eye of the beholder, necessarily involving an interaction of observer with environment. One can but pretend to a unique analysis of *the* information content of something, or *the* structure, since such notions are always relative to an agent's purposes and thought processes.

Yet information has gained an objective cachet by association with the physical concept of entropy. This was suggested by the formal resemblance of the equations involved in two distinct disciplines: communications theory and thermodynamics. Shannon information was defined as an analogue of physical entropy. Yet they are not categorically similar, for entropy measures randomness or disorder, while information informs.

As a measurable quantity, the information in communications originally referred to transmission rate and storage capacity for coded binary *messages*—implicitly reflecting the purposes of agents. But is entropy a message from nature? Though the entropy of information theory is associated with thermodynamic entropy, ultimately even the latter concept cannot be divorced from the purposes it serves. Nevertheless, contemporary physics seems to have ignored the caveats of E.T. Jaynes regarding the ambiguous nature of entropy.[21]

The strange fact that interference fringes occur only to the degree that the paths of particles remain unknown leads some to claim that the interference fringes appear only if the "path information" of the particle is "deleted" or not present in the first place.[22] This gives an unwarranted causal role to information. It describes a real phenomenon but in a misleading way. As demonstrated by "weak measurement" experiments, the occurrence of interference is not an all or nothing affair. Moreover, the notion that "path information" can be deleted or erased simply mixes metaphors, reflecting the influence of digital computation. We are accustomed to deleting information from *computer files*, which are designed to be reversible between two clearly defined logical states. It is nonsense when applied to real things. Quantum weirdness is not obviated by considering information to be a kind of caloric substance, which has causal power to prevent or allow interference, depending on its presence or absence. One can change the circumstance for individual particles (by opening or closing the slit, for example), but the information involved is simply one's knowledge of the state of affairs.

The association of a discrete ultimate physical structure (e.g., Planck units) with bits of digital information suggests to some that the universe is an information processing system. However, one cannot assume the "hardware" of the natural world to be well defined in the way that computer components are, nor physical process to be well defined like logical operations. These are but metaphors arising from current technology, based on the practical success of formal operations at simulating real processes. They do not grant *carte blanche* to view the universe as nothing but a computation or simulation, a virtual reality, a branch of mathematics, or a lengthy message from aliens! Such outlandish notions follow from adopting an idealist platform to begin with.

Some theorists propose to explain the discreteness of the quantum realm in terms of information.[23] That is, *nature* is quantized because *information* is quantized; hence, information is deemed the ontological substrate of nature. Such reasoning simply begs Wheeler's famous question: why the quantum? To say that the world is quantized because we have *defined* it to be so is hardly a physical answer to the mysteries of the micro realm.

The Game of Twenty Questions

John Wheeler's catchy aphorism *it from it* famously expresses the conviction that physical reality is "information-theoretic in origin."[24] What Wheeler calls *reality*, however, is what we have here been calling *knowledge*. Unfortunately, he does not distinguish generating information from generating reality itself.

Posing yes-or-no questions does, of course, generate digital information. Accordingly, the amount of information a message contains is the number of yes-no decisions required to convey it unequivocally. An artifact, being a product of definition, may indeed correspond to a message; its information content is then the number of yes-no decisions required to characterize it down to the last detail. But physical reality is not such a system or message. *Science* may be a message, with finite information content, but nature is not.

As Wheeler intimated, the information-theoretic view renders the scientific process rather like a party game, in which the science community asks a series of yes-no questions regarding a natural phenomenon. A question amounts to asking whether something falls in a given category or not, and the number of questions required to pin down the answer constitutes the information content. In the actual party game of Twenty Questions, contestants are to guess at a preestablished answer by a process of elimination. The series of questions converges to a known answer, and so terminates. To query *nature* in this way, however, may produce a nonterminating series; there are simply no preestablished answers regarding natural phenomena.

While the actual party game metaphorically represents classical realism, Wheeler modified it to represent the situation in quantum physics. In his version, the answer may change with each query so long as it is consistent with past answers. Yet this accommodation to the quantum realm only undermines the notion of a definite information content, since the answer in the modified game is a moving target, influenced by the questions. Furthermore, as in the classical realm, there is no limit to the number of possible questions or possible categories, so that the universe, or any part of it, cannot be characterized by a definite amount of information. Information is not generated by physical reality, but by the process of inquiry; it is stored and managed by the inquirers. It is supposed to correspond to real structure, but structure is ambiguous at any scale.

Conservation of Information, Holographic Principle, and Information Bounds

Is there a fundamental "law of conservation of information"[25] such that information cannot be lost—for example, within black holes? It does not follow that information must be conserved in the universe simply because it is conserved in some mathematical formalism. The need to rationalize the apparent "disappearance" of information, behind an epistemic wall such as the event horizon, only arises from reifying information in the first place. If information is not a kind of thing or substance, but reflects a cognitive state, there is no more (or less) ontological concern about its disappearance in black holes than there once was for its disappearance behind the Iron Curtain or the other side of the moon.

Reasoning about the properties of black holes has led to the further conclusion that our three-dimensional universe might "really" be a pattern of information on a two-dimensional surface, like a hologram. However, such a conclusion assumes from the outset a discrete ultimate structure of nature. (The dimensional reduction is actually gratuitous, since a limit—albeit bigger—would be implied even with three dimensions in a world already presumed to have an absolute discrete bottom.) The holographic paradigm is a vogue in physics, motivated by an idealist platform and the need to avoid infinities in mathematical calculation.[26] It argues for an ontological view of information as the fundamental constituent of matter (or an informational view of matter), by conflating two notions of information—one physical and one conceptual. To be sure, knowledge can be gained only through physical means, subject to energy limits, which can restrict knowledge. But the converse is not necessarily so: a limit on knowledge does not imply a limit on physical structure.

The notion of a finite information content, and therefore an *information bound*, rests on assumptions about possible degrees of freedom of a system. These depend on how the system is defined and partitioned—typically in terms of discrete entities, such as atoms, or their states. This fits nicely with digital computation and a definition of information that is expressly digital. With this definition and the physical assumption of discreteness, a calculable information bound appears inevitable. Applied to the universe as a whole, it then implies a limit on the total of information that can exist.

Yet one is free to ask awkward questions, such as: What is the relationship between information (entropy) and real structure in black holes? With what structures shall information be identified where structure is presumed destroyed or inaccessible? Insofar as internal structure is inaccessible, the apparent integrity of black holes resembles that of "elementary" particles. In some ways, the situation parallels the infamous *measurement problem*. But instead of a "collapse of the wave function" there is a collapse of the entropy when matter passes the event horizon.

While the *structure* of matter may be utterly torn apart inside a black hole, its gravitational influence, and hence its *mass*, presumably persists. Outside a black hole, where structure can exist and be identified with information, a greater mass normally corresponds to a greater number of particles and their possible states. Yet, inside, this identification cannot be made. Mass then becomes like thermodynamic variables, such as temperature and pressure, which disregard internal microstates.

Notes

1. Smolin, Lee. 2013. *Time Reborn*. Toronto: Alfred A. Knopf Canada.
2. Wigner, Eugene. 1960, "The Unreasonable Effectiveness of Mathematics in the Natural Sciences," *Communications on Pure and Applied Mathematics* 13:1–14.

3. A view early expounded by John Suart Mill.
4. Logic is more general than math. Note, for example, that the expression a=b is more general in logic (where a and b might denote distinct things and "equality" can have a variety of meanings) than in much of mathematics, where a and b denote numerical quantities.
5. For example, Baum, Eric. 2007. "A Working Hypothesis for General Intelligence." www.whatisthought.com/working.pdf. Proceedings of the 2007 Advances in Artificial General Intelligence conference. See also Baum. 2004. *What Is Thought?* Cambridge, MA: MIT Press.
6. Smolin, *Time Reborn*, p. 245.
7. Cf. Giere. Ronald N. 1999. *Science without Laws*. Chicago: University of Chicago Press, p. 92.
8. Turchin, Valentin. 1990. "Cybernetics and Philosophy." Eighth World Organisation of Systems and Cybernetics, New York.
9. Powers, Jonathan. 1982. *Philosophy and the New Physics*. London: Methuen, p. 140.
10. Schliesser, Eric. "Spinoza and the Philosophy of Science: Mathematics, Motion, and Being." PhilPapers. http://philpapers.org/rec/SCHSAT-23.
11. For example, Stephan Boltzmann's, Wien's and Planck's laws, and de Broglie's wave equation are held by some authors to be of this type. See: Caruso, F. and R. M. Xavier. 1997. "Space Dimensionality: What Can We Learn from Stellar Radiation and from the Mössbauer Effect?" www.academia.edu. .
12. Hartle, James B. 1997. "Sources of Predictability." arxiv.org/pdf/gr-qc/9701027. On the other hand, the scientific image of nature depends observationally upon the transparency of space, which may not have been so at an early epoch of the universe, and upon receiving light from distant galaxies, which may not be so in the far future. See: Krauss, Lawrence M. 2012. *A Universe from Nothing*. New York: Free Press.
13. Barrow, John D. 1991. *Theories of Everything: The Quest for Ultimate Explanation*. New York: Fawcett/Balantine, p. 50.
14. Reasoning that the world must be such as to allow living observers. See http://en.wikipedia.org/wiki/Anthropic_principle.
15. That is, causal factors not presently included in quantum theory. See http://en.wikipedia.org/wiki/Hidden_variable_theory.
16. Penrose, Roger. 1989. *The Emperor's New Mind*, Oxford: Oxford University Press, p. 403.
17. Two objects are qualitatively identical if they share all their state-independent properties (that is, if they share a common definition); they are numerically identical if they share their state-dependent properties as well.
18. Teller, P. 1998. "Quantum Mechanics and Haecceities." In *Interpreting Bodies: Classical and Quantum Objects in Modern Physics*, edited by Elena Castellani. Princeton, NJ: Princeton University Press, p. 116.
19. Ibid, pp. 119–20.
20. A quantity of information as defined in the study of signal transmission and related to the concept of entropy in physics. See: http://en.wikipedia.org/wiki/Information_theory.
21. See Jaynes, E. T. 1965. "Gibbs vs Boltzmann Entropies." *American Journal of Physics*, 33:396–98. See also Gull, S. F. 1989. "Some Misconceptions about Entropy." www.ucl.ac.uk/~ucesjph/reality/entropy/text.html.

22. Brukner, Č., and A. Zeilinger. 2005. "Quantum Physics as a Science of Information." In *Quo Vadis Quantum Mechanics?*, edited by A. Elitzur, S. Dolev, and N. Kolenda, Berlin: Springer, p. 48.
23. Ibid, p. 59.
24. Wheeler, John A. 1990. "Information, Physics, Quantum: The Search for Links." In *Complexity, Entropy, and the Physics of Information*, edited by W. Zurek. Boston: Addison-Wesley. More formally, "Reality arises in the last analysis from the posing of yes-no questions and the registering of equipment-evoked responses." This expresses a particular metaphysical stance.
25. As proclaimed by Leonard Susskind. See Susskind, Leonard. 2008. *The Black Hole War*, New York: Little Brown.
26. See Sorkin, Rafael D. 2005. "Ten Theses on Black Hole Entropy." http://arxiv.org/abs/hep-th/0504037.

Part Two

The World Remade

6

Consciousness and Its Consequences

"Only philosophy can presume to enunciate interesting answers to senseless questions."
—Enrico Bellone[1]

"Let those who want to call themselves philosophers bear the risk to their mental health that comes from thinking too much about free will."
—John Earman[2]

Which Mind-Body Problem?

Stub your toe and you feel pain. The pain seems to be a very different sort of thing than the stone that caused it. Stones and toes are physical things, but pain is an experience. What is the relation between physical reality and one's experience *of* that reality? Traditionally, philosophers talk in terms of *body* and *mind*—or *the physical* and *the mental*. The mental must be physically embodied somehow; yet the physical is a concept in someone's mind. This mutuality makes for a certain amount of tail-chasing confusion, rendering any definitive understanding of the mind-body relationship elusive. Some people try to explain the mental in physical terms, others vice versa. Whether materialist or idealist, those favoring one or the other have usually begun by assuming what they hope to prove. In other words, the so-called mind-body problem is a classic philosophical dilemma.

More broadly speaking, however, the relation between experience and reality is *the* human dilemma, touching every domain of our existence. It is the nitty-gritty of being mortal, vulnerable, physical, and social creatures that happen to have feelings and desires. The conscious self balks at its natural embodiment. Our ambivalence about being physical at all is one motivation for the flight from nature into idealization and culture that characterizes the human animal. Culture attempts to overcome the "natural condition" and assimilate nature to human preferences, agendas, and definitions, accounting in part for the technological urban world we live in.[3] Science is a particular manifestation of this cultural effort to substitute a humanly made world for the natural one. In this broader sense, the mind-body problem is not just an academic brainteaser about how conceptually to reconcile incompatible ontological categories. Rather, it bears the urgent concern of how to deal existentially, emotionally, ethically, legally, and scientifically with aspects of life that are unpalatable to a creature well aware of its condition.

At the deepest level, such concerns are more about suffering than intellectual challenge. They are often approached through religion, since suffering remains a constant of life in spite of scientific, technological, and social advance. Of course, such issues have been addressed by psychology, anthropology, and politics as well. The narrow slice that philosophy has sectioned for itself, however, tends to regard only the technical dilemma, which philosopher David Chalmers has famously called "the hard problem of consciousness."[4] It is ironic that this conundrum has been relegated to the pages of academic journals when its broader implications lie at the center of the human story, as the very force behind culture and history, reflecting the essence of what it is to be human. It poses the fundamental question of what we are; it forces us to consider, in particular, whether we are nature's puppets or godlike creators.

While our civilization may have grown hedonistic and over-subjectified in some ways, under the aegis of science it is also highly suspicious of subjectivity. We hardly trust our senses, our bodies and the messages they give us, the intentions of others, nor our own intuitions and common sense. We rely on experts for everything—including our view of the natural world, what ails our bodies, and the value and interpretation of our own experience.

Personhood

The awareness of being aware creates a tension, reflected in thought and language: an opposition between *I* and *it*. This distinction is more fundamental than particular beliefs about the external world, such as what kinds of things exist and which among them are sentient. It radically polarizes the realm of possibility into subject and object. While one may catalogue the perceived world in terms of sentience and non-sentience, for example, the differences between these can nevertheless be described in terms of the behavior of "it." While I am aware of my own inner life as a subject, I may be reluctant to attribute this to anything I can regard as an object, even an apparently sentient one.

Mind-body dualism involves the radical difference between the first-person and third-person *points of view*—a distinction reflected directly in many languages. Because of one's own conscious experience of being a subject, with a sentient body, one may suspect the possible experience of various objects, including other human bodies. This cannot be verified, however, since one's brain is hooked up uniquely to the object that is one's own body. The subject can look upon the world only from this unique perspective and not literally from the perspective of any other thing.

Among the possible experiences that claim our attention, some we hold to be external objects. And among these, some we hold to be subjects like ourselves. This does not free them from the category of "object," but gives them an ambivalent status, as witnessed by the human willingness to treat other persons as mere things. This ambiguity has worked both ways, so that what we presently regard as inanimate objects have sometimes been regarded as personal agents. That is, the distinction between subject and object was not always so clear as we hold it to be today.

To regard the natural world "objectively" means considering it to be impersonal, consisting of objects rather than agents active in the sense that human subjects are. Nature can be reduced to our mental constructs but it is not supposed to have a mentality of its own. One strategy for thus objectifying the natural world, stripping it of inherent powers, has been to consider it deterministic. The irony there is that if nature is deterministic and we are part of it, then we too are determined. On the other hand, if determinism is an urban myth (as I believe I have shown) then human freedom remains at least conceivable. We shall argue for the autonomy of nature, to which human autonomy is inextricably tied.

In addition to the natural dichotomy of *I* and *it*, the self-consciousness of a social being implies another dichotomy as well: *I* and *thou*. Verifiable or not, one naturally imagines some other bodies to be animated by the same sort of inner life as oneself. Other *persons*, along with other nonhuman agents, are thus understood to be active agents with free will like oneself, in contrast to the passivity of inanimate things.

As social creatures, human beings have significant relationships to one another. While one cannot *have* (or prove) another's experience, it is socially correct to assume that each other person has his or her own experience and is a free agent endowed with conscious awareness, in the same way one deems oneself to be. We have incorporated this category in language—in the "second person"—as well as in concepts of religion, psychology, and law.

Objectivity

The subject-object distinction is key to the evolutionary significance of "objectivity." The first-person point of view coincides with the organism's survival interests and programming; it is charged with "values," and focuses on objects significantly imbued with them. The behavior of the organism and its point of view fundamentally reflect these interests. In contrast, the ideal of objectivity is to be free from the idiosyncrasies of a point of view identified with the organism, and free from compulsory adherence to particular values. It is to be *dis*interested. This too has survival value, and is the basis of scientific objectivity; it serves the human organism to be able to override its built-in programming.

Self-consciousness, we have seen, is the ability to bracket experience *as* subjective. The evolutionary advantage this confers lies in the power to challenge the overconfidence we have in our perceptions, to question the compulsive self-evidence that attends value-laden judgments and feeling states in particular. Bracketing may be understood as a redefinition of the contents of experience. These are then to be understood as constructs made by the self, and under the self's power, rather than as objects found in experience, let alone found in the world. Paradoxically, this enables a more objective stance.

Bracketing serves at once to question the status of "givens," to assert responsibility for cognitive acts, and to appropriate to human agency aspects of the world presented in experience. Through conscious ideals, as well as through the natural bracketing process of self-consciousness, to some extent we have learned how to override our genetic and social programming. Religion facilitates this through faith in ethical standards and a concern for "higher things." Science does so through third-person description: the intersubjective world is the focus, not the anecdotal experience of the observer. Not without irony, objectivity itself represents a *value*, which may also be bracketed. Yet making it explicit has no place in first-order science since the very mention of values would call the pretense of objectivity into question.

The Consciousness Problem for Science

The very existence of consciousness, or subjective experience, poses a problem for first-order science, which in general deals in objectivist (if not truly objective) accounts. The problem of consciousness is not new, of course, yet seems to remain off limits to a mechanist approach. To paraphrase Leibniz, no amount of climbing about in the machinery of the brain explains one's experience of the color blue, for example, or why the desire to move one's finger leads to the finger moving. While there has never been (and perhaps cannot be) a definitive and universally accepted solution to this conundrum, periodic restatement of it gives at least an illusion of progress.

There has been genuine progress, of course, in the scientific study of the brain and of cognition. Drawing on computational metaphors, representational theories of mind attempt to explore the organism's role in perceptual processing and cognition. However, this is often at the price of considering the mind/brain as a self-contained system, and of considering cognition only from a third-person point of view. On the other hand, some modern cognitive approaches shun the notion of internal representation, to emphasize instead embodied interaction with environment. However, these approaches also tend toward implicitly behavioral descriptions, thereby also failing to come to terms with the problem posed by experience. Others apply a phenomenalist approach, though limited to the human subject and providing only a hand-waving solution to the problem of consciousness. There remains not only a problem posed by consciousness itself, but also an ongoing challenge to address it in a way that reconciles first and third-person points of view and divergent philosophical camps.

At root of these challenges is the ambiguity between objectively real things in the world and subjective experiences *of* them. This ambiguity is an inevitable outcome of reflexive consciousness, to which the subject-object split is endemic. It is not only a product of language, of European cultural history, or of philosophic tradition. The very nature of reflexive mind divides thought about itself, making the problem of consciousness doubly hard. It cannot be overcome by a one-sided approach in which reflexivity or subjectivity plays no part. Since even cognition can be treated either first-personally or third-personally, *point of view* must play an explicit role. At the very least, a reasonably complete theory of mind must hold that cognition involves a joint contribution—of organism and environment—to behavior *and* to subjective experience.

While a function of language, point of view reflects the psychological tension of an "explanatory gap" between subject and object.[5] In the scientific worldview, the problem posed by consciousness is the very existence of a first-person perspective in the context of third-person description. Attempts may be made to reduce this category dissonance—for example, by explaining subjective experience in terms of neural activity, in turn reducible to chemistry and physics. Perhaps because of the widespread influence of the computer, experience is typically considered an output of a cognitive system that is implicitly conceived in third-person terms. However, such maneuvers merely sidestep the dissonance. Scientific reductionism can explain *behavior* in terms of neural activity (or even computation), but does not readily explain *experience*. It does not close the explanatory gap nor solve the hard problem. On the other hand, it is perhaps because scientists and philosophers are committed to certain kinds of reductive answer that the problem is so hard.

Apart from the challenges of a scientific theory of mind, scientific narrative in general embraces third-person description, ignoring point of view as a factor relevant either to agency in the natural world or to the theorist's agency in creating the scientific portrait of nature. This has two consequences. First, causal agency is considered from the perspective of the scientist in terms of things or processes without a point of view of their own. This creates a skewed vision of the organism and perhaps of nature at large. No account of an organism's behavior, let alone of its inner life, can afford to ignore the process by which it establishes its own point of view. Since human beings are organisms, understanding that process is also key to solving the mystery posed by consciousness.

Secondly, first-order science is concerned with the material world, not with theory and methodology, which are generally left to philosophy of science to explicate. Of course, in order to understand the results of experiment or observation, the physics of the apparatus (and of the intervening causal processes involved in observation) must be considered along with the physics of the system under study. This much can be encompassed within first-order science using third-person description.

The observer is part of a comprehensive system that includes the apparatus of measurement, the information-carrying medium, and the system the information is held to be about. But nowhere is there provision in first-order theory to understand the relation between "mental" and "physical" or between first-person and third-person perspectives. Everything remains implicitly external and alien to the observer as conscious subject. Nor, in first-order theory, is the theorist's role *qua* subject considered—that is, there is generally no place for reflection on theory making itself, on methodology, or on implicit assumptions and biases involved both in theory and in experiment.

The exclusion of the observer in classical physics had to be qualified in the early twentieth century since it led to the problems that gave rise to relativity and quantum physics. Nor could this exclusion persist in view of the obvious contradiction of an observer outside the universe or the complementary problem of identifying the role of the observer in quantum phenomena. The program to ignore consciousness as a topic for study, which dominated psychology for most of the twentieth century, led to an undue focus on animal behavior (of rats, in particular), and perhaps to the depressing view that human consciousness simply does not matter. It endorsed a view of human being and of technology that could permit total war and a vision of nature as an inert mechanism—incapable of self-organization, let alone sentience, but highly suitable as a "natural resource."

Though we are accustomed to think of a yawning gulf between mind and matter, between persons and things, what the natural world and other persons have in common is that they are alike unknowns, similarly full of surprises. Even so, we regard most of the universe as inanimate, lacking any active ability to elude human goals of mastery. If it does so elude us, it is only because our net of definitions and concepts is provisional and too crude. While nature and its creatures are perhaps no longer viewed as worthy opponents, climate change may be restoring a sense of nature as a moving target that adapts to our influences. It can be argued that nature has every right to claim our attention in the second person: *as an unknown and unpredictable other.*

Notes

1. Bellone, Enrico. 1982. *A World on Paper: Studies on the Second Scientific Revolution.* Cambridge, MA: MIT Press.
2. Earman, John. 1986. *A Primer on Determinism.* Dordrecht, NL: D. Reidel.
3. See my earlier work: Bruiger, Dan. 2006. *Second Nature: The Man-Made World of Idealism, Technology, and Power.* Victoria, BC: Trafford/Left Field Press.
4. Chalmers, D. 1995 "Facing Up to the Problem of Consciousness." *Journal of Consciousness Studies* 2:200–19.
5. Levine, J. 1983 "Materialsim and Qualia: The Explanatory Gap." *Pacific Philosophical Quarterly*, 64:354–61.

7

What It Is Like to Be an Intentional System[1]

> *"The nature of things is more securely and naturally deduced from their operations one upon another than upon our senses."*
> —Isaac Newton

> *"If it should someday be proven that electrons are conscious, I would still want to know how they get their news."*
> —David Wick[2]

Embodiment and Representation

Embodiment is not merely physical presence, but a relationship with the world. The modern embodied cognition paradigm rightly emphasizes the organism's interaction with its environment. It acknowledges that neither the behavior nor the subjectivity of an organism can be accounted for by models that consider the nervous system to be self-contained or to merely process information that is passively received. For in accounts that lack the context of embodiment, there is nothing to show why an abstract and self-contained information-processing system should be motivated, have values or directives to govern its behavior, or have a point of view of its own, let alone why it should experience the world as real and external, imbued with phenomenal qualities. Hence, there is nothing to show how consciousness can arise within "inert" matter—even through self-organization. The origin of an organism's cognitive premises cannot be accounted for without an appeal to its embodied evolutionary context, which provides the reasons for its reasons.

Even so, there is something missing in such accounts, which remain in effect third-person descriptions. What is missing is the *organism's* point of view. From the observer's third-person standpoint, one never quite closes the gap with subjective experience, the first person. Even considering interaction with environment, such an analysis never makes it clear how bustling neurons produce the greenness of the color green, or how the feelings that imbue values, pains, and pleasures motivate actions. Even theories that emphasize the organism's active role do so by describing the *behavior* of an integrated *system*, whether this is the organism's molar behavior in regard to an environment or its detailed neural processing. The mystery remains that there is such a thing as "first person" at all. What is needed to comprehend this mystery is a concept of internal communication and agency that accommodates point of view.

The "Psychologist's Fallacy"

An observer may view the organism as adapting either to an environment or to its own state. Nevertheless, regardless of how (or, indeed, whether) an organism perceives its environment, one is tempted to view its actions as taking place upon what *we* perceive to be that environment or what we perceive to be its state. It is similarly understandable to identify the structure and functioning of organisms according to human categories, definitions, and purposes. William James dubbed this effect the "psychologist's fallacy." The very nature of the organism, however, is to be *self*-defining, to have its own priorities; it is only incidentally an object of human definition and study. While a machine or other artifact exhibits the priorities of its designers, we see from our human vantage point that the organism exhibits its own priorities, derived through an evolutionary and developmental history of interactions with environments consisting significantly of other players. This is the embodied basis of cognition and of the organism's point of view.

If the organism has sense organs, it responds to changes in their state through activity that restores a preferred pattern of sensory input. Even an organism without dedicated sense organs responds to changes of its own chemistry, attempting to restore a preferred state in ways that either prove successfully adaptive or not. With or without a sensory interface, the organism is, so to speak, flying by instrument, responding to feedback. The single cell and the human brain sealed within the cranium face the same challenge: to respond to changes of its state in ways that favor or permit its continuing life.

From an observer's point of view, of course, there *is* a world external to the cell wall, the skin, or the skull. Yet the challenge to the organism does not necessarily entail modeling an external world. The organism that does possess a nervous system may find it useful to engage in neural activity that the observer *takes* to represent or model an external environment. The observer may conceive such modeling in third-person terms, even as a relationship of input to output correlated by some algorithm. Such a third-person description is all very well for third-person observation focusing on the external world. But it fails when one tries to grasp how it is that one's *own* modeling process (algorithm) comes alive as personal experience, imbued with phenomenal qualities. This is the hard problem. Once we have admitted the possibility of experience at all, the situation is no different whether we are considering our own experience or the possible experience of other creatures or even machines. One way or another, the challenge is to understand the logic of the system's cognitive self-programming—*from its own point of view*. Indeed, the challenge is to understand how it creates that point of view. The handicap that stands in the way is the inevitable fact that, as observers, we can only occupy our own human and personal point of view.

The "Hard Problem" Made Slightly Easier

How can we explain conscious experience scientifically? How to bridge the gap between mind and body? First, we must acknowledge that the human brain *normally* bridges this gap: it is a physical system that does produce experience. Our task is to understand how it does this. Attempts to explain this in causal terms fail because they do not consider the organism's own point of view. Therefore, I propose that the brain bridges the gap through processes that are intentional as well as causal, as a form of internal communication. While signal transmission within the nervous system is a matter of *causal* connections, such as found by an external observer, the meaning to the organism itself rides upon connections *made* by, for, and within the organism. They are thus also *intentional* connections. It is this internal meaning that constitutes the organism's own point of view.

Signals are physical but involve meaning. The brain-body's communication with itself takes place through physical channels, but these embody intentional connections. The relation between intentional connections and causal connections is *like* the relation between the meaning of what you are now reading and the paper and ink (or electronic pixels) that physically convey it. Though there would be no communication without the text, the meaning requires you and me to create it. Similarly, the brain creates its own meanings, which ride on nervous activity as part of a world with which to interact.

Experience as Internal Communication

Embodiment involves a relationship of an *agent* with the world, whether that relationship is established through the present interaction of the individual or the interaction of the kind over generations of natural selection. While the behavior of organisms without nervous systems may not involve internal representation (nor even sensory information, but only direct interaction with an environment), the mediation of some form of internal modeling seems necessary for the cognitive functioning of nervous systems, and certainly for experience. While this does not imply a one-to-one correspondence of the model with external reality, it does suggest agency and mapping procedures that may be regarded as internal communication. Though neural mappings may be spatially arranged in the brain for reasons of economy, like all communication they are by nature symbolic, and not in principle isomorphic, spatial, or graphic. While such a mapping is not an image in a literal sense (no Cartesian theater), this does not preclude an internal agent that uses the information represented. Such internal communication, moreover, is *for the organism itself*, not for the benefit of an external observer. The information cannot be divorced from the organism's purposes and point of view. Information earmarked by an external observer is information for that observer, who is a separate agent with priorities and point of view distinct from those of the organism observed or any subagency within it.

Experience is correlated with behavior in obvious instances such as pain, which is a sensation involving a response. In that case, at least, behavior and experience have a common origin and meaning within the organism as an intentional agent. The flow of events can plausibly be described either in physical or mental terms as two descriptions containing the same information.

Information in general is both the message and the medium. On the one hand, it is knowledge of external events, meaning transmitted symbolically through intentional connections. Equally, information is a continuation of the events themselves, transmitted physically through causal chains of influence that pass both ways through the boundaries of the organism. Description formulated in terms of intentionality is thus potentially able to bridge the gulf between first- and third-person perspectives, because every intentional agent that exists physically is also a causal system. Simply put, every mind has (is) a body.[3]

Causal description concerns events in a physical system. In contrast, intentional description concerns steps in a *logical* system, like the moves in a game of chess, the instructions comprising a computer program, or the script of a stage play. It is a series of actions that are intended or prescribed. (In the light of mechanism, we think of computer programs as self-contained algorithms; but they are also expressions of someone's intentions.) "Intention," however, must not be restricted to *conscious* intention of human beings.[4] The important point is that *connections* may be at once causal and intentional; indeed, they must be in the case of embodied minds. Intentional connections within an organism are both neurological and "logical"—while not necessarily rational or conscious.

While intentional description may explicitly take a third-person form, it implies a first-person point of view. It is an analysis of the intentions and point of view of an agent, as human observers can understand them. We may describe intentional behavior in terms of logic circuits or flow charts, for instance, which seem to have an objective existence, apprehended in third-person terms. Yet these tools refer ultimately to the designer's intention; they do not merely provide physical description, such as would be the case if the circuit were described only in terms of soldered wires. Moreover, one must be careful to differentiate between human categories and purposes and those that might be the organism's own. We are looking over the creature's shoulder, so to speak, as we would over a programmer's shoulder in trying to grasp the logic of the program. The difference is that one cannot take for granted that such "logic" bears any resemblance to the considerations of a human programmer.

From a formal point of view, intentionality involves symbolic operations gratuitous in themselves. The cognitive system of an organism can be understood as a potentially formalizable logical system, which can be "interpreted" as referring to the real world. Intentional connection takes place in logical order rather than temporal order: the if/then of syllogism rather than cause/effect in time and space. Of course, such connections must also be physically embodied if they are to exist in the material world. The whole point is their correlation, just as the meanings of words are carried on corresponding physical sounds or signs.

Just as squiggles on a page can come alive as a story and mathematical symbols can represent actual relations between things in the world, mind *is* the semantic sense that the brain makes of its own representations, which it communicates to itself through neurological signals. The development of the underlying connectivity is guided by natural selection through which a physical system comes to be an intentional system, acquiring a point of view of its own through its long evolutionary history. It is through this history of interactions with the world that it makes sense of its relations to the world, registered in internal representations that *matter to itself.* Motivation is the source of meaning, and motivation comes from embodied participation in an evolutionary contest. It is thus that the "preestablished harmony" between experience and world is established.

Intelligence is an elaborate form of adaptive behavior. Treating it as a disembodied phenomenon is the main reason for the failure of classical artificial intelligence. No computer (yet) stands in an embodied relationship to an environment that provides not only its input, but also the *significance to itself* of that input and, hence, a motivation for developing output that insures its continued existence. It seems that such a relationship can only be established bottom up through an evolutionary contest. The premises of the embodied mind are not arbitrary. They are not programmed from outside or imposed from on high, but inhere as values implied in genetic fitness.

Qualities

Yet even intentional description remains essentially a behavioral third-person account, telling us nothing of first-person experience. There still persists an explanatory gap so that we are left wondering how a physical system can *have experience.* What, after all, do the felt qualities of experience—the greenness of trees, sweetness of sugar, or burning of pain—have to do with the blob of gray stuff inside one's head? It seems too enormous a leap from slimy brain cells to the personal spectacle of sensations and feelings, which one is ill prepared to make even in the age of neuroscience and computers.

Partly at fault may be our simplistic idealizations of matter in tandem with equally idealized notions of selfhood. The mechanical models with which Leibniz was familiar were hopelessly simple. A modern computer is unfathomably more complex than a clock; yet even it is nowhere nearly as complex as the lowliest organism—a single cell. Above all, however, we are victims of our metaphors, being ourselves tied in the Gordian knot we attempt to unravel. One cannot find a strictly causal (let alone mechanistic) explanation for consciousness, when the understanding of causality precludes intentionality, meaning, agency, significance, and—of course—consciousness. These are the very things, encompassed in Aristotle's thought, that were rejected as "subjective" by the natural philosophers who set the future course of science. The dualist outlook, in which the observer is a fly on the wall outside a material system, precludes an understanding of the observer herself. An objectivist outlook, no matter how sophisticated, will never by itself close the gap between mind and matter.

I have suggested that the self-luminous character of *qualities* in sensation (the redness of red, the hurtfulness of pain) arises in much the way that the meanings of language do. Sounds are made to carry meaning as words through a constructive and symbolic process. (Similarly, algebraic symbols gain numerical significance by convention and through usage.) Phenomenal qualities that "emerge" in consciousness are comparable to intelligible meanings that emerge through the babble of spoken syllables, or through the squiggles on a written page. This creation of meaning is an active process involving internal communication, largely about interaction with an environment full of consequence for the organism.

While qualities in general may thus be understood in terms of intentional connectivity, the significance of *specific* qualities must be understood in terms of the evolutionary advantages of *particular* connections, with their particular real-world references. As with the etymology of words, the evolutionary origin of particular "qualia" is a topic meriting study. The fact that healthy vegetation, for example, is generally experienced as some shade of *green* cannot be arbitrary. The sensation of greenness (unlike the word) is not merely a linguistic convention, but a convention of neuro-logical organization, with the force of long genetic precedent. Indeed, the human cognitive system gradually adapts to distorting colored lenses or filters in such a way that experience of verdant foliage—to pursue the example—is restored to a normal greenness.[5] Such a "convention" is backed by evolutionary history, just as words come to be imbued with stable referents through a history of usage in the world. The sensation of greenness is inherently different than the sensation of redness precisely because of the real-world things it refers to in our evolutionary history, from which it cannot be arbitrarily dissociated.[6]

In any language, including mathematics, the meaning of each symbol is conventional. It is posited by intention, but ultimately refers beyond itself. This applies as well to the internal communications of the organism. The greenness of the color green, the hurtfulness of pain, the spaciousness of space, the solidity of objects, even the realness and externality of the world—all these qualities reflect such posited meanings, established by fiat within the organism, conveyed in the "language of the senses," yet shaped by properties of the real world.

Notes

1. Philosopher Daniel Dennett refers to systems whose behavior can be attributed to beliefs and desires as *intentional systems*. While this includes human minds, it is more generally a way of looking at natural and artificial systems to understand behavior that appears to involve purpose.
2. Wick, David. 1995. *The Infamous Boundary: Seven Decades of Heresy in Quantum Physics*. New York: Springer-Verlag, p. 180.
3. The reverse, of course, is hardly true. Not every physical system is a mind—a point sometimes overlooked by post-humanists who read into any complex structure a potential home for mind.
4. Intentionality should not be confused with intention, in the ordinary sense, but clearly they are related. In logic, "intension" specifies what properties apply to objects while "extension" specifies which objects they apply to. Clearly these logical relationships are akin to the intentionality of mind, on the one hand, and the spatial extension of physical objects and causal relationships, on the other. An agent ascribing properties to something *intends* in both senses of the word.
5. Neitz, J., J. Carroll, Y. Yamauchi, M. Neitz, and D. R. Williams. "Color Perception Is Mediated by a Plastic Neural Mechanism That Is Adjustable in Adults." 2002. *Neuron* 35:783–92.
6. This is why inverted-spectrum arguments do not work. The "inverted spectrum" is the idea that two people could share color vocabulary and discriminations yet have systematically different corresponding experiences.

8

The Rebellion against Nature

> "Man seeks to form for himself... a simplified and lucid image of the world, and so to overcome the world of experience by striving to replace it to some extent by this image."
> —Albert Einstein[1]

> "We were not given dominion over the earth; our forebears earned it in their long nightmarish struggle against creatures far stronger, swifter, and better armed than themselves, when the terror of being ripped apart and devoured was never farther away than the darkness beyond."
> —Barbara Ehrenreich[2]

The Human Predicament

The human creature has a foot in each of two worlds. We are magnificently good at carrying on life; biologically our species is a great success. Yet underneath the momentum of daily life in many ways the human situation is unsettling. We know that we are going to die, that civilizations rise and fall, that even the planet will one day no longer support life. We know that misfortune and suffering are always possible for us personally and always actual for many around the world. We see that technological change is not necessarily social progress. In spite of godlike pretensions, we suspect that we are selfish beings at heart, over which nature holds a deeper sway than morality and law. We fear that we are at the mercy of forces we don't understand or control, whether natural or man-made. We are assailed by doubts on all sides.

I believe that such doubts are a driving force of human culture. The very fact of self-consciousness motivates everything from artistry to war, from religion to science, from walled cities to the wall-to-wall technology of death stars envisioned in science fiction.[3] Uncertainty has been present from the beginning, when culture was but comforting stories told around the campfire. In contrast to the eternity and boundlessness intimated in consciousness, we are haunted by the realization that we are finite, vulnerable, mortal—and perhaps meaningless—creatures. As science expands empowering knowledge, paradoxically it also deepens this realization of the awesome indifference and enormity of the universe as a presence beyond human ken or control.

Perhaps the philosophy of mechanism grew out of this sense of the impersonal and indifferent life of the cosmos—even as a kind of psychological defense against the expanding view of the world emerging in the Renaissance. Explicitly viewing the world as a machine was a ploy to turn the tables on the alien vastness of the universe, by reducing it to a device of human conception and proportion. Long before, however, the self-conscious creature had intuitively realized that it could and must define its own world. Hope lay in the realms of its own inner life rather than in the natural world that holds it hostage. While people could shape the external world to conform in limited ways to their desires, even more importantly they could redefine the world in their own terms, narrating their own story. To begin with, they could personalize the forces impinging on them, asserting their will through supplication or magic. They could later *de*personalize these forces for similar reasons—gaining an upper hand by usurping their power.

Mankind would have to adapt to the new worlds of its own making. If it wished to live apart from nature, in a bubble of culture, reason and technology, it would have to remake itself as an artificial creature. This corresponds, I propose, to a deep yearning for self-generation implicit in human consciousness. If so, the human being is foremost an aspiring and idealizing creature who flees mortality, limitation, embodiment, the corruption of time, and determination by nature, in pursuit of freedom within an ideal world of its own design. The primordial hope behind all culture is for a humanly conceived and controlled environment that is also a repository of meaning beyond the death and decay of the individual. This is the burden of expectation implicit in any civilization, against which its failings are ultimately measured. It is a burden inherited by science.

Natural Despair

Like all else in nature, the human phenomenon is an example of something that happens because it *can* under the right conditions. Yet conditions can change, so that life in the protective bubble of our biosphere exists on a precarious ledge in space and time. The astronomical record testifies that most planets are unsuitable for life. The geological record testifies to mass extinctions of once-successful creatures. The living landscape is ever shifting, so that the present diversity of life represents far less than 1 percent of all the species that ever lived.

Nature generally does not go back to the drawing board, but cobbles changes onto existing structures. This means that in many cases nature's designs are far from perfect—that is, from the ideal perspective of a human engineer. While the system as a whole works, it does not necessarily work to the advantage of a given individual. Nature designs statistically, opportunistically, and does not care as much as we like for the fate of its individuals. Of course, a human being *is* such an individual whose life and well-being depend crucially on details of natural design. This dependency cannot help but be a source of disaffection, and a motivation to substitute for indifferent nature a world ordered to human taste.

There are many reasons for despair over the natural condition, beginning with mortality, hunger, and disease. Whatever else we are, we are physical organisms dependent on the natural world; embodiment is the general condition for our mental life. The interests of the individual organism coincide generally with those of the soma, while the rules of the game are dictated by the interests of genes. The interests of the individual human personality, however, are potentially independent of both. We suffer in awareness of our dependency, therefore, to the degree that our actual goals, visions, values, and experiences are at odds with the system of nature, thwarted by the body, or determined by forces outside ourselves. From the point of view of a self-aware consciousness trapped within it, the system of nature is a senseless game with little regard for individual welfare or happiness. Yet survival requires commitment to this game, which means obeying the dictates of those genes. The dawning human ego, though knowing nothing of genes, would not have gone to the sacrifice willingly, but found ways to resist and struggle against nature's tyranny.

Psychologically, we have the peculiar ability to experience the body at once as a foreign object and as personal property. The body is both slave and master. We are bound to it by every need, beginning with the need for air. We suppose it to be the servant of personal will, but it has a mind of its own. It is understandable, then, that human consciousness would turn not only against the natural world, as prison to the body, but against the body itself as prison to consciousness. One is pitted against one's *own* body, in a struggle for control, and against the corporality of the world at large. The possibility of transcendence is suggested by the interior subjective space that seems to provide a refuge both from the body and from the external world. Withdrawal into the realm of mind or spirit—the realm of the Ideal—expresses alienation from the physical and its powers over us.

The notion of causality was first modeled on the direct experience of will that comes from successfully intending the actions of one's own body.[4] This personalized understanding of cause also served to assimilate intimidating natural events to the human realm and to the conscious control of will—if only through magical thinking. What we now take to be impersonal forces could formerly be propitiated in the way that people can be. The *impersonal* understanding of cause as "natural" is but an afterthought, when identification of scientific laws and technology permitted a more effective means to trump nature.

Along with civilization generally, religion and science alike can be viewed as strategies to cope with the deeply embedded perception of nature as indifferent, alien or cruel, threatening human sensibilities from without and from within. In Christian traditions, this perception is mollified by considering nature the rational creation of a provident God. In the scientific tradition, nature is disarmed through considering it inert, passive, malleable, subject to the rule of governing law, manipulation through technology, and redefinition within theory. So to speak, man has turned the tables on nature by regarding it as a vanquished enemy, which now even requires our aid and protection in some environmental plan of reconstruction. Nature local to this planet is now humanity's victim, rather than the victimizer it once was for most of our species' existence. The idea of nature as benign or neutral—or as a stable backdrop to the human drama—is a romantic fancy reflecting the tamed nature we know in parks and gardens, the paper and pencil nature we know in science, and the entitlement we assume as masters of technology. Even so, the marvels of civilization and our very existence remain every bit as contingent as the marginal Homo sapiens was at the dawn of prehistory.

This natural state of dependency only enhances the significance of transcendence as an ingrained freedom of human consciousness (even when, ironically, interpreting this freedom as submission to divine will). Mankind has *always* demonstrated the initiative to remove from nature's grasp and define its own way. That is what culture is. We render ourselves human through conscious idealization, even re-creating human nature itself. We are the creature that remakes itself—an identity that coexists uneasily with our evolutionary programming. It *is* a serious problem that modern environments bear little resemblance to our formative ancestral one, so that our "innate values" are poorly matched to contemporary life. Such values can lead us back to tribalism, far more dangerous in the modern context, and to the destruction of the cosmopolitan ideal of a united humanity. But they could also lead us forward to a renewed localism and a more humane scale of civilization. The nineteenth-century decline of religion simply invited political and economic ideologies to take its place as galvanizers of faith and moral action. The failures of such secularist programs are no doubt partly responsible for religion's resurgence in the twentieth century. These shortcomings must be assessed—along with those of religion—in the light of the imperative to take conscious charge of human destiny. Biology and evolutionary history may inform the human future, but they do not entirely fix it. Ours is the freedom to experiment—against biology, and now *with* biology—which always includes the freedom to fail.

The Domination of Nature

The domination of nature can be understood as a reaction to domination *by* nature. To compensate for the early experience of victimization, in which they were literal prey, human beings set forth on a general path of dominance, to turn the tables on nature by learning the aggressive role of predator.[5] The whole of culture, in fact, may be viewed as a movement to take active charge of the human condition, substituting man-made terms for vulnerability in the wild, re-creating the found as the made. The deep mark left upon the human psyche by its early vulnerability motivated people to ensure, in so many words, that it would never happen again.

Whether waged against animals, people, or nature at large, war is a way to proactively disarm death and injury by rendering them deliberate. Hunting, war, and human and animal sacrifice were once highly ritualized collective enterprises. Their *symbolic* role was to sacralize death—and, thus, to mitigate its sting. Ritualized practices are significant precisely because they are *not* events in nature, but within a human world redefined in ideal terms. It is not just death that is transfigured, but the whole of embodied vulnerability. The numinous aura with which these activities were once charged mirrors the deadly power the natural world holds over human life—the same seriousness we have seen to underlie natural realism. Culture is mankind's answer to human dependency—and, therefore, to nature's ominous *in*dependence.

Realities of the found world are thereby translated into narrative, assimilated to the fabricated human world.

The Human Realm

Establishing the distinctly human identity and realm is not only about denying our animal heritage. More positively, it is about affirming a godlike power and transcendence, a spiritual standing. Whereas the animal must suffer its passive existence in a found world, human beings create their own world—and with it their own suffering. The human drama is set apart from nature because it is defined in terms of will and idea, as opposed to material cause and effect. The cultural significance of science and technology is not simply their utility for survival, to shield people from natural contingency. They also signify entitlement to dominion over nature and constitute a revised version of the natural world.

Man is the creature who seeks to transcend natural limits, to be located outside nature and time, in a world of human definition and devising. Accordingly, the scientist creates a vision of nature as though from outside (the external observer), from a timeless perspective (the "block universe"), located in sheer abstractions ("configuration space"). Yet a brief walk in a rainforest reveals the extent to which the laws and concepts of physics do not characterize the bedlam of nature, but only the idealized world of the physicist.

In fact, the comforts of certainty can be found *only* in such an idealized world. The corollary is that scientific knowledge, to the degree it is certain, is *not* knowledge of nature at all, but of scientific concepts.[6] Hence, scientific knowledge reflects the structure and needs of the human psyche as much as it reflects the reality of nature. While correlation must be presumed between theory and reality, it can only be tested through experimental set-ups that are themselves conceptual and literal artifacts. Effectively, scientific knowledge directly concerns artifacts of one sort or another; only indirectly does it concern nature.

Mankind appropriates nature's powers through technology, attempting to remove itself from the natural arena altogether. This was first accomplished by redefining reality in language and building human environments. Then, by reconstructing nature in scientific concepts and technological artifacts, even restructuring matter itself. Finally, by controlling life processes, breaking free of the gravitational bonds of the planet, and now by taking up residence in cyberspace, leaving behind even the notion of reality.

The Masculine Hero

While the heroic quest of culture is to create the specifically human world, this has been conspicuously a male project, reflecting a masculine version of humanness. Having been defined by men, heroism is traditionally a masculine preserve. The conquest of nature, of civilizations, and of intellectual and spiritual realms, has been largely a male enterprise. The story of man's relationship to nature is in many ways the story of men's relationship to the feminine.

From antiquity, men associated the active causal principle in nature with masculine agency, spirit, will, mind, or law, in distinction to passive substance, which they identified with woman. Given their inferiority in his view, Aristotle considered the very existence of women an enigma. His antifeminism was matched by that of Augustine in the medieval Christian world. Western intellectual traditions first developed in the celibate and misogynist culture of learning of the cathedral schools and universities—a world literally without women.[7] Many of the great philosophers and scientists—among them Newton and Boyle—were celibate or solitary, in a long tradition repudiating women and family life. Many intellectual culture heroes have been stoic figures, loners forming few close personal ties and relatively indifferent to bodily needs, riding off into the sunset of history.[8] Even for others, the attraction of science surely includes the high quest for pure knowledge and the possibility to make an enduring contribution to human culture. Modern science is at once competitive and cooperative—very much a team effort—which perhaps enhances the sense of importance, contribution, and participation in a grand undertaking.

The heroic quest of Greek idealism had been taken up in Christendom through the figure of Christ—the male literally deified. The emulation of Christ eventually led to the heroic figure of the Magus and his descendants, the explorers and natural philosophers. Renaissance scientists, inventors, artists, mapmakers and explorers became culture heroes of a new sort. The Europeans' very ability to measure, to survey, to make detailed maps and representations, conferred in their minds the right of ownership over "new" lands, whose native inhabitants had no such notions. Knowledge is power, and a map of any sort displays condensed knowledge, conferring power over the territory it represents. Symbolic appropriation leads to literal appropriation; as far as the conquistadors were concerned, the map *was* the territory.[9] An echo of this attitude carries over into modern science, where theoretical models are taken more or less as realities.

On the shoulders of religion, humanism invoked a Promethean concept of man as a secular being whose reason liberated him from dependency on divine revelation, bringing him a huge step closer to the core project of self-generation, ready to rebel against Heaven and Earth, to dominate the planet.[10] Heroism is a quest for meaning, for a significant place within the scheme of things. Perhaps one reason that science (and humanism generally) is able to compete with religion is that science offers an ongoing heroic participation in the communal salvation. Though subject to interpretation and debate, religious truth is in principle complete and fixed, while scientific truth is arguably open ended.

The power of abstraction, fostered especially by writing and then printing, enabled men to turn the natural order upside down and re-create it in a male image. A notion of creativity distinct from the procreativity of women fostered a shift in what is considered real—from the evident material presence of nature, in which women play a special role and hold an elevated place, to an invisible, immaterial abstraction, of which men are the masters. It is largely men who created the philosophies and religions of transcendence. It is still largely males who create the dominant planetary culture, which "heroically" seeks to transcend the unsavory facts of embodiment.

One may speculate that gender conditions the sort of concepts that may arise in science as in culture generally. A masculine ethos "naturally" conceives forces, for example, acting to overcome the inertia of passive matter.[11] Even more fundamentally, the fact that a living body can act upon other bodies, but also be acted upon by them, leads inevitably to the duality of active/passive, subject/object, and thus to an apparent choice between dominating nature or being subject to it. If the dominant ethos had been more feminine, perhaps the very concept of force—or efficient cause—might have played a lesser role in physics or been more subordinate to properties inhering in matter.

The hero is the culture bringer, the rebel against the natural condition, whose mission is to establish the human world. To the extent that culture is invented, its forms can appear arbitrary. We call the narratives of other cultures *myths*, in contrast to those of our own, in which we take consolation. On one level, faith is necessary to the heroic endeavor, which is framed within the terms of the group with which one identifies. Without cultural premises, one lacks a context of belonging in which to have individual identity; and without such identity one is merely an animal that will die. While one hopes that such identity will transcend mortal embodiment, the human is the animal that *knows* it will die. Yet if it can live with this knowledge, it should be able also to live with the knowledge that the cultural forms it believes in are but makeshift barricades against the natural condition.

Heroic Science

Whereas the animal happens to affect its surroundings, the human *intends* to. Whereas the animal remains within the world of nature, the human forges a world of ideas and attempts an ideal world that corresponds to them. Through technology, the narrative of science brings creative invention to found reality. As an active dialogue with nature, science answers nature's existential challenge as much as it solicits nature's answers to its questions. For it does not only inquire into what exists, but boldly replaces natural ambiguities with its own definite constructs. It actively substitutes a symbolic household for the found world. Physics in particular, with its ambition to unify fundamental forces within a monumental framework, reflects the heroic quest that underlies all culture—to humanize the face of the unknown.

As a form of cultural heroics, science is a quest for ultimate truth and the ultimate constituents of reality—or at least for a satisfying story concerning the natural world. It is a secular creation story that must be acceptable to reason and compatible with experience. It must also capture the imagination. Bacon's vision of the social role of science was to restore humanity to its rightful place before the biblical fall, promising salvation through the use of technology for social benefit. The heroic vision of the Enlightenment was that society could be grounded in rational, scientific principles, based on the true interrelationships of things rather than upon superstition, religious faith, or wild speculation. Generations of later scientists saw nature in terms that inspired industrialization and technologies to probe the universe at the largest and smallest scales, in search of a completed vision.

Yet the darker side of this quest is control: the "violence lurking in all positive and communicable knowledge."[12] Monumental heroics have been nowhere more ironic than in the "science" of economics. The ideal of centralized control of economies has proven as counterproductive as top-down artificial intelligence. On the other hand, "free" market capitalism has been self-defeating from the beginning, since it tends toward monopoly rather than competition, social disparity and tension rather than fraternity and democracy. The excesses of nineteenth-century robber-baron capitalism had led to ideas of centralized state control, based on principles of top-down management. State control appealed to capitalist as well as Marxist governments, to regulate market fluctuations and mitigate their disastrous social consequences. As demonstrated by the fall of communism, such control was far more challenging than imagined.[13] This realization led to less public regulation and the present globalist program—in which regulation has simply migrated from public control to the hands of an economic elite.

The whole point of economics is prediction and control. Yet an economy is not a fixed system whose dynamics can be studied and controlled from outside, but an ever-changing game of many mutually influencing players, including economists. Economics aims to manage "natural resources"—which means the planet itself. The disasters of economic prediction should be a lesson for natural science. For nature is a complex economy of sorts, involving mutually influencing factors, including "economists" who approach it with simplistic schemes.

Notes

1. Albert Einstein, ca. 1918, quoted in Holton, Gerald. 1996. *Thematic Origins of Scientific Thought.* Cambridge, MA: Harvard University Press, p. 395.
2. Ehrenreich, Barbara. 1997. *Blood Rites: Origins and History of the Passion of War.* New York: Metropolitan Books.
3. See Bruiger, Dan. 2006. *Second Nature: The Man-Made World of Idealism, Technology, and Power.* Victoria, BC: Left Field Press/Trafford.
4. Piaget, Jean. 1977. *La construction du réel chez l'enfant.* Lonay, Switzerland: Delachaux & Niestlé, p. 10 and chapter 3.
5. Ehrenreich, *Blood Rites*, p. 47.
6. Cf. Einstein's famous quip: "In so far as the propositions of mathematics refer to reality they are not certain; and in so far as they are certain they do not refer to reality." Einstein, Albert. 1954. *Ideas and Opinions.* New York: Crown, p. 233.
7. Noble, David F. 1992. *A World Without Women.* New York: Alfred A. Knopf, pp. 136–137.
8. Anthony Storr gives Newton, Descartes, Locke, Hobbes, Pascal, Spinoza, Kant, Leibniz, Schopenhauer, Nietzsche, Kierkegaard, and Wittgenstein as examples. Storr, Anthony. 1988. *Solitude.* London: Flamingo/HarperCollins.
9. Cayley, David. 2010. "The Origins of the Modern Public." *Ideas.* CBC radio broadcasts. www.cbc.ca/radio/ideas. John Dee, for example, was court mathematician to Elizabeth I. Like Galileo; he held that nature is mathematical and that those who can decode it carry the mandate of heaven.

10. Nasr, Seyyed Hossein. 1996. *Religion and the Order of Nature.* Oxford: Oxford University Press, p. 176.
11. Holton, Gerald. 1988. *Thematic Origins of Scientific Thought.* Cambridge, MA: Harvard University Press, p. 42.
12. Martin Heideggar, cited in Prigogine, Ilya, and Isabelle Stengers.1984. *Order Out of Chaos: Man's New Dialogue with Nature.* New York: Bantam.
13. Of course, the demise of the Soviet Union was not only due to top-heavy governance, mismanagement, and corruption but also to its military competition with the West, which drove it to excesses beyond its means.

9

The Ideal of Perfect Knowledge

> *"It is far better to predict without certainty than never to have predicted at all."*
> —Henri Poincaré[1]

> *"For physicists, the only good Demon is a dead Demon."*
> —H. C. von Baeyer[2]

The Value of Certainty

We come upon the natural world apparently as something found, to which we have faith that our knowledge corresponds. Yet it is essentially an unknown that does not readily give up its secrets. Just as the materials of a craft have properties that must be forcibly bent to the designer's will, so nature's inherent properties may resist the forms imposed on it by thought. One can only guess at the parts, structure, and organization of systems that one did not make in the first place! Nature's autonomy renders it elusive and implies that no description of it can be exhaustive. We have noted already that the only systems that *can* be exhaustively described are made, not found.

Natural autonomy casts a pall of uncertainty over all human enterprise, which can only unfold in the risky venue of nature. Pascal eloquently expressed the anxieties of his age on the cusp of the Scientific Revolution, turning to religion for certainty. His contemporary, Descartes, sought it rather in the possibility of deductive knowledge, which stood in contrast to the certainties of religious faith, mere probability, and opinion based on the testimony of the senses. Hobbes believed the "state of nature" to be chaos, in which the condition of uncertainty would be unbearable—a prime reason for civilization and the social contract. Laplace found certainty in mathematics and the philosophy of determinism, which became the pillar of classical physics. Modern thinkers turn to computation as an empowering tool, to model natural reality and as a model *for* it.

The sheer fact of subjectivity places us in a chronic state of doubt. If perception and knowledge are products of the self, as well as of the world, how can we know that they are true to reality, that the scientific vision of nature is reliable? The very autonomy of nature means that our accounts of it cannot be complete and may fail.

In daily life, within a world that seems essentially stable, one is not much concerned about such niceties, which are left to the philosopher or scientist. Like all else in our specialized world, inquiry into such matters is confided to experts, part of whose social responsibility is to give back a plausible story that answers fundamental questions, relieves anxieties, and enables society to carry on with business as usual. Yet even experts are prone to keep the personal implications of "deeper" issues at arm's length, by intellectualizing them and by containing their professional life within office hours.

Prizing certainty, however, we are scarcely satisfied merely to recognize limits of our capacity to know. Even in scientific thought, the admission that reality eludes us must be reframed as a property of reality itself. For instance, the uncertainty principle is often held to represent a qualitatively different kind of limit to knowledge than what one encounters in classical physics. It is cast in the light of an objective property of nature at the quantum level rather than as an epistemic failure. However, it can be shown that a similar principle, of the same mathematical form, applies even in the classical realm.[3]

The Idealized Observer: The Demons of Laplace and Maxwell

While perfect knowledge may not be possible, we hold it as an ideal. For Newton, there had been at least one omniscient observer: God. In nineteenth-century physics, the ideal of perfect knowledge was personified rather by the "demons" of Laplace and Maxwell. Like the evil genius imagined by Descartes, whose job (to systematically falsify one's perceptions) could require unlimited knowledge, these beings are hypothetical observers with extraordinary cognitive powers. They appear in thought experiments in which limitless knowledge is invoked to explore the limits of physical concepts.

Laplace's demon is a disembodied intelligence with access to infinitely detailed information about the states of all particles in the universe: an idealized observer with perfect knowledge of the initial conditions from which any future or past state could be calculated in a completely deterministic world.[4] Laplace's demon personified this Mephistophelian power for several generations of physicists. A deterministic world, after all, is highly desirable from the point of view of mathematical treatment, prediction, and control—provided only that one is not part of it! The fact remains, however, that most human experience cannot be treated mathematically at all. The success of science must be measured against an overwhelming background of questions that cannot be scientifically addressed.[5]

James Clerk Maxwell is best known for equations of electromagnetism bearing his name, from which the speed of light was predicted. As a mathematical theorist, he was also interested in statistical mechanics—the theory that explains the behavior of gasses in terms of the motions of molecules. This stood in contrast to thermodynamics, which describes large-scale properties such as heat flow, volume, pressure, and temperature. The second law of thermodynamics was a hot topic in the nineteenth century since it seemed to imply that the universe is dissipating toward an ultimate "heat death." It says, basically, that the overall amount of disorder in the universe cannot decrease, and that efforts to extract useful energy and create local order result in more overall disorder or unusable energy. There are no perpetual motion machines and no free lunches.

However, Maxwell was suspicious of attempts to prove the second law using statistical mechanics and devised a thought experiment to explore the issue. Imagine a container of gas with a central partition. A hole in this partition allows molecules to pass from one chamber to the other so that pressure and temperature would equalize over time. But imagine also a miniature helper inside who effortlessly operates a frictionless hatch placed over the hole in such a way as to admit only fast-moving molecules one at a time, one way into one chamber. Eventually this would create a higher temperature or pressure on one side than on the other. Work could be extracted from this difference in a repeatable cycle (e.g., by moving a piston)—apparently contradicting the second law. For more than a century, physicists have been refining and debating the implications of this conundrum, trying either to exorcise Maxwell's demon or to prove that the second law is not inviolable. It's still unclear whether the second law should be thought of as a matter of principle or as only statistically true on average, and how it might apply to the cosmos as a whole.

Unlike any real observer, Maxwell's demon is a disembodied intelligence, whose physical properties are negligible. Neither the demon nor the hatch have any mass or friction and require no energy to move. (In other words, the demon appears to defy a law of physics by cheating physics in the first place! The challenge is to understand how he gets away with it.) Maxwell's thought experiment not only explores concepts in thermodynamics but also points to general questions about the validity of idealization and the role of the idealized observer. This, in turn, reflects the awkwardness of our position as both physical and mental beings. Like these assorted demons, we are ambiguously participants and observers, standing at once inside and outside the system—and always hoping for a free lunch.

While "demons" were initially conceived to represent the ideal of perfect knowledge apart from physical factors and consequences, the possibility of cheating the second law can be conceived in strictly mechanical terms, bypassing "intelligence" per se. Yet in each version of the thought experiment, it seemed evident that the demon could not function without increasing the overall entropy, thus affirming the second law. At first, it was thought the entropy increase must be due to the measurement process. It was later shown that the increase of entropy comes from erasing rather than measuring or storing information. But the story has not ended there. Recent claims have been made that even erasure can be entropically costless.[6] Similarly, claims are now made that "information" can be converted to usable energy to do work.[7] Intuitively, however, a *molecule* can do work, but *knowledge* about it is not physical and cannot do work unless it is somehow physically embodied—for example, as another molecule in a memory system.

Self-Reference, Circularity, and the Problem of Cognitive Domains

Many people are familiar with Escher's images of impossible staircases and hands that draw themselves. These are graphic depictions of circular reasoning, often involving self-reference. Just as Escher's illusions make a logical impossibility seem materially plausible, so lack of a proper distinction between logical domains may obscure such reasoning.

I call this general situation the "problem of cognitive domains." It concerns the relationship between map and territory, when the only access to the territory is through the map. It involves a kind of epistemological cheating.[8] In such situations the notion that models can successively approximate reality cannot presume a direct *view* of what is modeled, only a better *performance*. That is, the fit of the map to the territory cannot be defined in terms of *accuracy* when no direct comparison is possible because there is access only to the map. The fit can then only be defined in terms of *adequacy*, which is ultimately a matter of evolutionary fitness.

Attempts to explain how the mind builds its picture of the external world, for example, often begin with the very picture of the world they attempt to explain. This beginning point may include such elements as photons, neurons and electrochemical processes. Through these the brain is presumed to construct an image—of photons, neurons, and electrochemical processes! We treat our image of the world as though it was the world itself, which is then recycled—in the bootstrapping style of the legendary Baron von Munchausen—as the cause of the image we experience![9] Thus, a neurologist might regard some neural process as causally responsible for her thoughts about (and even perception of)[10] that very neural process. The conclusion to be established is assumed as the premise; the domain of explanation is recycled as its own cause.

There are, of course, many instances of circularity in physical science that aren't about consciousness per se. For example, in reasoning backward to the big bang, one may wonder at the meaning of time in an environment where no time-keeping processes could exist. We use categories formed in the present era to try to grasp an era for which such categories may not be appropriate, just as we use categories appropriate to one scale that may not be for another.

Models are proposed to explain observed phenomena, which are then conceived in the terms of the model, as though that is what they "really" were all along. The fit between deductions from the model and observable phenomena is to be tested in experiment. But so conceiving the role of experiment already presumes one knows what the variables to be measured are. While a measurement can be made without explicit reference to theory, theory often specifies what is to be measured, and any interpretation of measurement must appeal to it. Yet since the experiment is a test of the theory, the whole thing bites its own tail.

Broadly speaking, the mechanist metaphor is projected back upon nature as the organizing principle behind the very life and consciousness that creates the concept of mechanism! (Yet far from being a reasonable model for understanding nature, mechanism is the very *opposite* of nature's organic nature.) Metaphors, paradigms, and theories can also feed forward by setting research agendas—creating, in effect, self-fulfilling prophecies. An example from biological science is how DNA-sequencing technology has shaped genetic research. The fact that drastic variations can be produced artificially by gene manipulation falsely suggests that all natural variations are due to genetic mutation. This is also an example of how the experimental setup can determine an outcome that does not match the natural reality, since many natural variations are more subtle and may be produced by other factors.[11]

Macro and Micro

To understand how macroscopic properties emerge from microscopic properties, one must sort out the relationships between distinct domains of description. How do liquidity and solidity as macroscopic properties, for example, arise from the behavior of molecules? Nobel laureate Robert Laughlin points out that physicists are accustomed to think about solidification in terms of the packing of little spheres, while atoms are not really such objects and lack even an identifiable location. Phase organization does not emerge from Newtonian concepts and objects but the other way around.[12]

It is sometimes said that the state of a quantum system becomes real only when it is measured.[13] However, "realness" becomes paradoxical as a property that may be acquired or lost through the intervention of an agent. Moreover, the problem posed by measurement appears in the macroscopic world as well: the classical equivalent of Schrödinger's cat is Berkeley's tree falling in the forest. One may choose to believe that macroscopic things are distinguished from microscopic ones by the fact they continue in their real state between sightings. But this conclusion is little more than circular reasoning. The best we can do is to shorten the interval between observations and assume continuity.

The quantum realm confounds ordinary expectations because we assume that notions based on familiar experience should apply in other realms, on other scales. Laws of physics predict which *kinds* of particles exist, for instance, but cannot identify a given particle as distinct from another. At the quantum scale, it seems there are not distinguishable individuals, only examples of kinds. Quantum phenomena are collective and statistical in essence, but this does not deter the temptation to conceive individual events as though they were macroscopic. The logic grows circular, however, when properties on the macroscopic scale are conceived to emerge from events on the microscopic scale, while the latter are implicitly conceived in macroscopic terms.[14]

In the macro realm, the observer is assumed to stand outside the system observed, which the observation does not unduly influence. Bohr refuted this for the quantum realm, stating that no such separation is possible and the interaction is unavoidably significant.[15] Yet even in the macroscopic realm the division between observer and observed is a matter of convention and the difference in the effects of observer interaction in the two realms is a matter of degree. As Bohr pointed out, there are not two worlds (or even two scales), only two approaches. The many issues surrounding "realism" in quantum physics arise largely because of our conventional division of the world into subject and object. This division is afforded on the macroscopic scale by the extreme asymmetry of size or energy between ordinary objects observed and the quanta of light through which they are observed.

The World as a Black Box

From the point of view of the scientific investigator, the physical world is a sort of black box, about whose content she speculates on the basis of inputs and outputs. An experimental situation is a well-defined version of some part or aspect of the world. It defines an input in order to observe an output, inviting speculation on the hidden processes by which the world transforms one data set into the other. These processes can be modeled with equations or computer programs (which have corresponding well-defined inputs). Nature, however, cannot be "opened" to view its contents "directly" (that is, apart from theorizing and experimentation). The *actual* box is the experimental setup, which in principle is defined in accord with the theory. The experiment is in fact a *substitute* for the natural phenomenon investigated. While it is ideally a perfect analogue, an indefinite number of alternative models might work to account for the input-output relations. Theory tells us how to think of the input, the output, and the apparatus. The experiment is supposed to determine to what extent nature corresponds to the model, but the interpretation of its outcome depends also on how well the experimental setup corresponds to the natural situation investigated.

The scientist is in a similar position as the brain sealed within the windowless skull, with laboratory equipment serving as sensors. If the brain seems a black box to investigators outside of it, the world outside the skull is equally a black box to the brain. The brain can only compare changes to its own inputs in order infer whatever is going on "out there." So much is true for the scientist as well, who can but compare instrumental readings that result from changed experimental settings.

There is an important difference, however. In the case of perception, we come literally to experience our cognitive processing *as* the world itself. We live in the model, so to speak, as though it were the world, seeing the world through it, as though it granted direct and transparent access. This "illusion" is a strategy of evolutionary history, without which we would probably not be here. It expresses the species' blind faith in its cognitive adequacy. Scientific models also come easily to be taken for what they model, but with the important difference that a wrong scientific theory is not usually lethal, whereas a wrong perceptual model easily could be.

Clearly, the brain is a survival tool, and in some sense science is too. Human evolutionary success seems to reflect not only the reliability of ordinary cognition but also a growing adequacy of thought and culture—including science—as a biological strategy, though, of course, the jury is still out on this question. Yet, one may ask, how exactly does mathematical modeling serve evolutionary fitness? What is the relationship, in general, between cognitive adequacy and truth? The fact that a wrong theory hasn't yet killed us doesn't make it true. It is tempting to think that technological advance is proof that the scientific portrait of nature grows ever more accurate over time, and therefore more adequate for human survival. However, the advance of technology does not in itself necessarily reflect a true or adequate picture of the world, let alone guarantee survival. The sole meaning that can be assigned to accuracy of the model is that its predictions accord with observations construed on basis of the model itself. And the only meaning that can be assigned to adequacy is that we do in fact survive, which could simply be a matter of luck. Strictly speaking, one is entitled to assert only that the scientific view of the world has not so far led to our extinction. On this footing, the minds of other extant creatures make an equal claim to represent reality adequately.

Multiple Causes

As Maxwell had noted, when theorists find a deviation in observed results from the predictions of theory, their first response may be to ascribe this to some extraneous perturbing cause rather than fault the theory. He also recognized that a new causal factor *could* be little more than shorthand for the aspects of reality not covered adequately by the theory.[16] Nevertheless, one should generally expect a given effect to have multiple causes, and causes to have multiple effects. The question of *which* factors are to be considered significant involves choices that affect how the system is analyzed from then on.[17]

Even the idea of a system of separately identifiable causes working together requires that factors must be isolated in the first place to be definitely identified (in which case, by definition, they are *not* working together). There may in fact be no such thing in nature as an isolated cause, and therefore no such thing as a group of such causes working in definable concert. The very notion of *factor* stems from the sort of linear analysis whose limitations a more holistic reasoning tries to overcome.

Notes

1. Poincaré, Henri. 1905. *Science and Hypothesis*. New York: Walter, p. 144.
2. von Baeyer, Hans Christian. 1999. *Warmth Disperses and Time Passes: The History of Heat*. New York: Modern Library, p. 145.
3. Hamming, R. W. 1980. "The Unreasonable Effectiveness of Mathematics." *American Mathematical Monthly* 87:81–90.
4. No wonder that Laplace had no need for the "god hypothesis," since he thought he could provide the mortal scientist with an omniscient view! To be fair, he did propose that (failing the possibility of perfect knowledge) the next best thing was probability, which he used to support his nebular hypothesis of the formation of the solar system.
5. Hamming, "The Unreasonable Effectiveness of Mathematics."
6. Hemmo, M., and O. Shenker. 2012. *The Road to Maxwell's Demon*. Cambridge, UK: Cambridge University Press.
7. For example, Toyabe, Shoichi, Takahiro Sagawa, Masahito Ueda, Eiro Muneyuki, and Masaki Sano. 2010. "Experimental Demonstration of Information-to-Energy Conversion and Validation of the Generalized Jarzynski Equality." *Nature Physics* 6:988–92.
8. Kenny, Vincent. 2009. "There's Nothing Like the Real Thing: Revisiting the Need for a Third-Order Cybernetics." *Constructivist Foundations* 4:103.
9. Baron von Munchausen famously saved himself from drowning by lifting himself out of the water by the scruff of his own pate, horse and all. Perhaps the notion of *bootstrapping* involves a similar reference to the days of horsemanship!
10. If she were performing her own brain surgery, for example.
11. Lewontin, Richard. 2000. *The Triple Helix*. Cambridge, MA: Harvard University Press, pp. 15 and 128.
12. Laughlin, Robert B. 2005. *A Different Universe: Reinventing Physics from the Bottom Down*. New York: Basic Books, pp. 31, 42, and 56.
13. e.g., Nadeau, Robert, and Menas Kafatos. 1999. *The Non-Local Universe: The New Physics and Matters of the Mind*. Oxford: Oxford University Press, p. 59.
14. See, for example, Landau, L. D., and E. M. Lifschitz. 1977. *Quantum Mechanics*. Oxford: Pergamon, quoted in Bell, John S. 1990. *Against Measurement*, *Physics World* 3:35: "Thus quantum mechanics occupies a very unusual place among physical theories: it contains classical mechanics as a limiting case, yet at the same time it requires this limiting case for its own formulation."

15. Bohr, Niels. 1961. *Atomic Theory and the Description of Nature.* Cambridge, UK: Cambridge University Press, pp. 53–54.
16. James Clerk Maxwell, cited in Morrison, Margaret. 2000. *Unifying Scientific Theories: Physical Concepts and Mathematical Structures.* Cambridge, UK: Cambridge University Press, p. 96.
17. Hut, Piet. "Ambiguity at the Roots of Precision." www.ids.ias.edu/~piet/publ/other/ambiguity.html.

10

The Scientific World

"Space is extremely well named."
—Bill Bryson

"Time is an illusion. Lunchtime doubly so."
—Douglas Adams

Introduction

Like all creatures, human beings thrive in stable environments. A world defined by human agreement at least promises to be more dependable than the unfathomable ways of nature. Since humanly invented systems are in principle knowable by definition, they represent a relatively secure knowledge. It is tempting to embrace such knowledge *because* it is dependable and readily accessible. Hence, one should not be surprised, for example, at the complacency of medieval scholars who preferred copying manuscripts to observing nature.

We live in culture and civilization, not in nature. The need for security goes far to explain this estrangement. It may also shed light on science as a human enterprise, insofar as it too is a quest for the certainties of a reconstructed version of nature. At the level of such a deep cultural directive, science merges with other cognitive modes, such as religion and metaphysical speculation. At that level, theorizing and mythologizing alike fulfill this need. Nor can we readily distinguish between pure knowledge and other goals, such as control of natural forces, social betterment through technology, and commercial exploitation of natural resources. For all these are aspects of the project to establish the human world.

We have noted that the epistemic situation of science parallels that of ordinary cognition. From a constructivist point of view, experiment and observation are formalized cognitive acts that create new appearances to save by means of theory.[1] The natural world does not unilaterally inform theory, which has also a creative aspect involving a community of subjects who vet theories through competition. The explanatory role of visual or mechanical models, regardless of their truth, attests to human creativity and the general significance of artifacts in culture. Models, like all artifacts, are important in their own right because they affirm our ability to construct a user-friendly version of nature.

While the natural world is ambiguous, concepts are what we specify them to be. Whereas events in the physical world are acts of nature, events in the scientific world are acts of human beings. Thought is inherently simplistic, whereas natural and even social realities are complex, tending to elude formulation and control.

The chaos of the human world, as much as the chaos of nature, may have inspired the early search for underlying unity, simple truths, and reliable procedures that could be widely agreed upon. The need for universally accepted rules of debate grew out of the very competitiveness and individualism of ancient Greek thought. The discovery of formal reasoning extended the art of rhetoric, which was as much a form of competitive sport as it was an earnest dialogue. Like their Greek forebears, Enlightenment thinkers turned to laws of reasoning for intellectual soundness and to laws of state for political security. The Royal Society may have been formed, in part, in reaction to the religious and political turmoil of the English Civil War and of the Reformation, more generally.

The ideal of individual thought and argumentation presupposes the cooperative aspects, ground rules, and social forms of the collectivity. The seventeenth-century institutionalization of science grew out of a sense of spiritual brotherhood that permeated the religious culture of Europe. The development of axiomatic systems and the certainties of logic depended on the ability to imaginatively share given assumptions and rules of thought: agreeing to disagree. These are the elements of a game, voluntarily and consensually embraced, allowing problems to be approached collectively with accepted rules. Yet assumptions and rules are at once enabling and limiting. In science too, the need for formal structuring can conflict with the need for novelty.

Problem Solving

Science is a search for problems as much as for solutions. Problem solving keeps scientists employed. Yet the kinds of problems pursued are shaped by past success. Scientists tend to pursue those problems that yield to their methods.[2] Moreover, a driving force behind burgeoning specialization may be the fact that scientific papers in a given field are now produced faster than they can be read.

Problem-solving activity provides a psychologically manageable zone in which one is suitably challenged and engaged without being overwhelmed. Such a comfort zone is bordered at one extreme by chaos and at the other by boredom. It is said that children need structure, but so does the adult mind. Culture in general provides structured bounds within which social play can reasonably be conducted. Various social forms and activities serve a cognitive function beyond their ludic aspect; conversely, cognition structures the babble of sensory input in the way that games structure play. Like myth, scientific narratives function in part to "spare us from the complexity of the world and shield us from its randomness."[3] In addition to nature's inherent structuring, the very game-like structuring of science serves to reduce cognitive dissonance.

Any game with appeal has a playing field somewhere in the comfort zone. This might correspond to a physical space, as in a sport; but in essence it is abstract, a product of definition. The game is a formalism, with well-defined elements, rules, and goals. This is both its advantage and its disadvantage. Formalism can be problematic when it determines the actual environment in which one moves. While it is liberating to create order, order can also be entrapping. One then lives and perceives *through* the game, in the *world* of the game. This makes other games, other rules, more difficult to imagine. More insidiously, the world of the game displaces the natural world when the game is the study of nature. No doubt this reflects the sort of trade-off culture involves generally. Just as most people now live in cities, so scientists—though some may claim otherwise—live in the world of science, not in nature.[4] For many, their work environment is literally in cramped offices or basement laboratories. Nature itself becomes hearsay, even literally a museum piece.

While conceptual models are needed to mediate a scientific concept of nature, they subtly eclipse nature itself as the object of study. In the extreme, science then becomes a neoscholastic study of texts and commentaries upon commentaries, upstaging the realities that are supposed to be its objects. Formalization involves a shift from empirical to theoretical truths, from found to made, with the risk that science becomes knowledge merely of its own constructs. More than streamlining is involved in the attempt to model complex phenomena in simplified, idealized, and formal—most often mathematical—terms. Apart from quantification, there is a qualitative shift from the real natural thing, as the object of study, to a conceptual artifact. One sees this in the increasing reliance on computer models and simulations as a basic new genre of research.

In part, the Scientific Revolution was a revolt against a kind of speculation driven less by observation than by the need for tidy principles, categories, and relations. Ironically, this need—for a self-contained and self-evident system—remains the core of theory, as well as of theology. That scientists are not immune to this need is attested by the fact that many of the early scientists were theologians. Many participated in the rearguard challenge to reconcile biblical doctrine with the discoveries of the new world and the new sciences. Even today the drive for theoretical coherence and unity maintains an uneasy balance with the requirements of empiricism.

The Closure of First-Order Science

"First-order" description is an account of events in the physical world. While this restriction obviously serves to keep science within proper bounds, and is responsible for much of its success, it also represents a limit on the kind and the terms of reflection that may be undertaken. One symptom of the exclusion of self-reference is the tendency to recycle a domain of description to serve as its own rationale (the problem of cognitive domains). This results from using a system that does not accommodate reflexivity in a world in which reflexivity is an unavoidable and essential aspect.

The physical world may be cognized differently by various actual or theoretical organisms—as well, of course, as by different people or cultures. However, we normally take the physical world as *we* cognize it to be the *actual* world. At least the modern *human* way of telling the story of the world serves as a default reference on which to judge possible other accounts. It is commonly assumed that mathematical theories of physics fill this role of default or ground-base truth because mathematics speaks a language underlying all possible cognition. But that article of faith simply defers to another: the transcendent reality and universality of mathematics, which (whatever else it is) is a cognitive construct made within recorded human history.

The mind is endowed with a capacity to recognize invariants and gestalts of experience, and to disregard dissonant details. This boon comes at a cost, since it is possible to be overzealous about it. Generalization and categorization can wrongly dismiss exceptions that might lead to recognizing deeper patterns, types, or laws, or that might have a significance not yet understood. Emergent pattern and signal-to-noise ratio are to some extent discretionary, testifying that the world does not have a unique structure. If it *did* have a unique structure, one would expect there to exist only one standard and optimal cognitive system, not the diversity we see.[5]

The notion of alternative universes suggests physical possibilities other than what we actually perceive. This places a standard observer imaginatively within a "landscape" of possible worlds. A typical first-order account might focus on historical events leading to the type of universe harboring observers such as ourselves. Cognition then simply adapts to the conditions of such a universe and plays no physical role in setting them.[6] Any other kind of role is not discussed, so that despite reference to observers one remains within the bounds of first-order description. There is no consideration, for instance, of a landscape of possible observers.

The recognition of mathematical possibility outside physical actuality has often led to the discovery of new phenomena. Perhaps, similarly, the recognition of broader "cognitive possibility" outside our particular cognition could also lead to new discovery. Physics is highly mathematical, and it is reasonable for physicists to believe in the guidance of mathematics. Yet if physics is a form of cognition, then it is also reasonable to believe in the guidance of cognitive theory, evolutionary psychology, and what might be termed theoretical epistemology. Mathematics by itself does not usually take us outside the closure of first-order description.

Another effect of first-order closure involves the belief that we should be able to understand complex, non-linear processes in terms of familiar deterministic models. Self-organizing processes, for example, are presumed to be extensions of known mechanisms, with the reductionist promise that living matter can ultimately be understood in terms of chemistry and physics. Yet we still do not fully understand the differences between living and nonliving things. The search for the simple linear relationships behind such models developed for historical more than logical reasons: that's what could be done with the intellectual resources available. With the digital computer, we now possess vastly improved resources that enable the study of non-linear processes and greater complexity.

Searching in the Light

While nature imposes limits on our knowledge, it is essential to understand limits we ourselves bring to its study, which include an all-too-human proclivity to settle for readily accessible constructs.

This brings to mind a Nasrudin story: When asked by a passerby what he was doing down on his hands and knees under a street lamp late at night, Nasrudin answered that he was looking for his misplaced house key. When asked where he might have lost it, Nasrudin replied that he had no idea. When asked why, in that case, he was looking *here*, he replied that here, at least, there is some light! Understandably, scientists prefer not to be utterly stymied by things they do not yet, and perhaps may never fully, understand. While it makes common sense, the strategy of looking where it is convenient and feasible to look, or where technology encourages us to look, may ultimately be no more logical than Nasrudin's.

Astronomers literally search in the light. The amazing evolution of telescopes has been driven by the need to collect ever more of it. What once appeared as faint fuzzy patches in our own Milky Way were only in the last century understood to be "island universes." What appeared as isolated galaxies, in the photographic plates made in an earlier generation of telescopes, have been revealed by more modern equipment to be connected by vast filaments torn from each other gravitationally. What once appeared as an empty void between them is now seen to be more of an active plenum—at least within galaxy clusters.

Yet the most astounding realization has been that what can be "seen" at all (in the entire electromagnetic spectrum, not just in visible light) amounts to only a tiny fraction of the apparent mass of the universe. Our very concept of knowledge is based on the visual sense and on what the eyes or their extensions can detect. Our epistemology is based on the principle that no relative motion can exceed the speed of light. Our ontology (the standard model of particle physics) is founded on what was historically a theory of the emission and absorption of light. Are we to assume that the other 95 percent or more of the universe fits into the categories we have devised for the miniscule visible part?

Granted that scientific understanding is unavoidably provisional, it is difficult to take seriously the prospect of a definitive theory. Science is more plausible as a system of conceptual supports for technical recipes that tell us how to manipulate physical processes and materials to advantage. From a user's point of view, such a system would ideally be consistent, unified, and logical: a deductive system. Hence, for better and for worse, we tend to define scientific knowledge in terms of such systems. In the next few chapters, we shall further explore some of the underlying reasons for this.

A Scientific Establishment

We have already looked at scientific process through the lens of a judicial metaphor. Scientists propose and debate various models, as though arguing before a jury or legislative body. Experiment alone is rarely decisive, but serves rather as evidence for persuasive argument. In other words, science is a creative activity, not just a mechanical procedure.

It is commonly held that science, unlike the arts, represents a growing body of objective knowledge. This fits well with our modern idea of progress. However, what actually accumulates is *data*, which always remain open to new or revised interpretations. Evidence grows (if we preserve it), while theories come and go. If one assumes from the outset that there is a knowable truth of the matter on which the evidence must converge, then it makes sense to believe that theories get closer to that truth with time. However, the evidence often is not consistent or does not converge as well as hoped. Nevertheless, scientific bravado, abetted by the media, often presents the verdict as beyond a reasonable doubt, suggesting that the evidence unambiguously *dictates* it. This belief, however, fails to take full responsibility for the scientific due process that is subject to human limits and motivations. Even in science, what is "reasonable" doubt is a matter of opinion.

While the science profession (the "bar," so to speak) establishes official tolerances for error, what may go unexamined are many tacit assumptions the "court" fails to discuss or even recognize in the pressure to reach a verdict. What is admitted as evidence is already guided by suspicions of guilt. Unlike in law, however, dissenting opinion is systematically ignored or forgotten, regarded as error, and filtered out of classrooms and textbooks, which treat decisions of the scientific community as accumulating fact. (Laws and court decisions do accumulate, and affect current practice, but dissenting opinions are also often preserved and may be later used as new arguments.) Another way to put this is that law is more self-reflective than science; it incorporates its own history, rather than leaving that to outsiders. Unlike human law, the revisionist approach of science makes the laws of nature seem falsely simple, self-evident, and historically inevitable.[7] It also creates the impression that nature itself can be axiomatized, with finality, according to the lights of the current generation of scientists.

What science appears to debate is less the truth of nature than which theory an expert elite should embrace at a given moment in history. The laws of nature are then decrees of the scientific establishment, echoing a time when they were considered divine decrees, or else edicts of the religious establishment. In one sense, this "secularization" represents laudable progress. It is, after all, the role of subjectivity to claim responsibility for perception and for belief founded upon it. However, first-order science treads a thin line between actually taking that responsibility and simply proclaiming its findings to be established truth.

The monopoly of ideas, like the monopoly of wealth, implies a class system. Until relatively recently these coincided: only the rich could afford education and had the leisure to exchange ideas. As a product of the Reformation, science more or less usurped the position of the Church it displaced, claiming semi-divine authority for its dispensations. An academic elite now underwrites the ability and right of an economic elite to manage the world economy, just as the Church once endorsed the authority of monarchs. Such elites overlap and tend to be similarly insulated from both nature and social reality. As in medieval times, ivory towers are vulnerable to intellectual fashions, promoted to the rest of the world as eternal verities.[8] Nowadays, the domination of intellectual space by a few should be as troubling as the unequal distribution of wealth—more, perhaps, because it is harder to mitigate through social policy.[9] The scientific establishment is effectively a gated community. The problem with the control of knowledge by an expert elite is that it may obstruct knowledge that it does not already lay claim to. While physics is explicitly concerned with the unification of forces in nature, perhaps the advantages and disadvantages of its own monolithic structure should be of more concern. While science often prides itself on the universality of its institutions and (mathematical) language, there may be a down side to the lack of diversity that implies. Monoculture of any sort is less robust than diversity.

While consensus is socially desirable, scientists know that it is no guarantee of truth. Yet it obviously plays an important role in the scientific world, as in other social realms. Consensus seeking, however, can become a political game, a competition for the highest number of citations in journals. This often has an economic motivation, since scientists are competing for research funds; publication is then as much about attracting money and prestige as alerting colleagues to new finds.[10] Senior researchers typically get published in the more prestigious journals, so that their preferred interpretation in a controversy tends to prevail.[11]

The Influence of Digital Technology

Information technology is the latest expression of the mechanist philosophy. We are informed by what actually surrounds us, and digital technology constitutes a large part of the modern environment—perhaps more so for scientists. Younger scientists, growing up with information technologies, are understandably conditioned to regard information as an *ontological reality*, and information processing as a *natural* process. The ubiquity, power, and advantages of digital technology invite us to believe that nature is literally digital at a fundamental level or that the universe itself is somehow a computation, consisting literally of information.[12] Such unwarranted leaps are but the latest development in a long tradition that posits mathematical abstraction as the reality behind appearances, and a tradition that embraces the latest technology as the model for understanding nature. It is no more plausible, however, that physical reality should happen to correspond to the technology of one century than of another.

An obvious and defining advantage of digital technology is the fidelity of information transmission within the system. A digital program faithfully yields the same result, given the same input, each time it is run: perfect determinism. However, the internal self-consistency of digital systems has nothing to do with fidelity to the real world. In that regard, the computer's output is no better than the input, and this advantage stops at the interface with the world outside.

As a research tool, it is often cheaper and more convenient to run computer simulations than experiments involving elaborate equipment. Yet the computer is not only a new means to study nature but is also the latest metaphor for nature itself. This is reflected in ideas like the "mathematical universe hypothesis," the "it from bit" philosophy, and the notion that physical reality fundamentally consists of digital information. While there is great practical advantage in digital computation—and in thinking in terms of information—it is a simple category error to assert that the physical world literally *consists* of information, mathematics, or computation. Such claims ignore the fact that information requires a sender and a receiver and computation requires a programmer or user. Moreover, transitions between physical states are classically conceived to be continuous, reversible, and modeled by differential equations. Transitions between states in a classical digital computer are discontinuous by definition, and generally confined to one direction; intermediate states are not defined at all. Most real numbers are non-computable, which is an inconvenience for a science based on the continuum (differential equations). However, while physics *could* be expressed in terms of a discrete mathematics, this would reduce it to a deductive system, with all that does not fit into its Procrustean bed left undefined.[13]

Pedagogy

Children tend to believe in those invisible agents (such as germs) that adults are seen to endorse. In contrast to chimpanzees, they tend to slavishly copy all steps involved in a complex learning task. This suggests that steps that intuitively seem to them inefficient or counterintuitive, or that they cannot understand, are nevertheless taken on faith to be important.[14] The truism—that we look to our role models for guidance concerning what is real and how to deal with it—has possible consequences for the ways scientists approach physical reality, especially in realms beyond the scale of ordinary experience or beyond the ideas and methods of the present generation of luminaries and teachers.

Circularity and closure occur not only in the practice of science but also in its teaching. Textbooks typically approach science as though it followed a logical rather than historical development. While this method has advantages, it presents the current state of knowledge (or some version twenty years out of date) as eternal truth. Each generation's ideas are thus enshrined as absolute and final. A loosely axiomatic approach creates the impression that nature itself can be finally axiomatized in just the way that it appears to the current generation of scientists. Entities currently posited by theory are reified as actually existing in a way that circularly justifies the theory that posits them. They gain the weight of acceptance through a social and pedagogical process that is not strictly logical.[15] Some esoteric areas of theoretical physics promote their byzantine creations as mainstream science.[16] Many cosmological speculations rely on tenuous threads of reasoning about very thin evidence, yet are presented not as speculation but as the latest portrait of how the world really is. The manufactured account of nature becomes a narrative that erases its tracks, removing all trace of historical process, dispute, or context. The "definitive" version is recycled as the object of scientific inquiry—the nature that allegedly is found.

The Rebellion against Science

Should education teach us how to make a living but not how to live? Should science tell us how to manipulate nature but not how to coexist with it? What should take up the slack between such divergent concerns? As the twentieth century gained momentum, liberal intellectuals expected that religion would continue its apparent slow decline, that secular institutions would continue to gain ground, displacing ritual and superstition, and that society would at last become fully "rational," as the Enlightenment fathers had hoped. This, of course, has hardly been the case. Two devastating world wars and the absurd threat of "mutual assured destruction" cast serious doubt on reasoned guidance of the species' destiny. We have seen the economic promise of the modern age wither under economic globalism and free trade. Effectively, our touted secular institutions have only served to magnify social disparities, much as though this had been their purpose all along.[17]

Interference by power and money in the affairs of ordinary people, at home and abroad, has understandably led to resentment and resistance as well as impoverishment for many. More broadly, it leads to revolts against the values of "reason," the emptiness of consumer materialism and its impacts upon the environment, and the narrowness of the secular vision endorsed by science. This, I believe, is one cause of religious resurgence around the world and of antiscientific sentiment. The scientific narrative may be adequate to sustain technological and economic "progress" for a time, but it does not inspire society to reach the potential of the humanist and religious ideals that motivated it in the first place. Rather, science has become sidetracked within specializations and seduced by technological accomplishments instead of aligning itself forcefully behind such ideals. There are many reasons for this, ranging from the tacit agreement of the Royal Society to stay out of politics to the corporate interests behind genetic research.

The scientific establishment now claims hegemony over society's view of the material world. While nowadays this is contested mainly by religious fundamentalists, at various times there has been more widespread secular resistance as well. The internal struggle of society—between reason, represented by science, and more vital forces in reaction—forms part of an overarching contest between opposing archetypes in western thought, mirroring the antagonism between mind and body. Any such dialectical struggle must result in cycles of fashion, for neither aspect of human nature will ultimately triumph over the other. Rather, there will always be a precarious balance, with one tending to dominate in a given generation or epoch.

"Romantic" movements have rebelled from time to time against the rule of reason and its treatment of nature, including the movement in the first half of the nineteenth century that was largely a reaction to the Industrial Revolution. Sometimes there is a strong political dimension, as in the case of socialism and Marxism as reactions to the capitalist excesses associated with industrialization. The rise of fascism after the First World War reacted to the punitive "rationality" of postwar reparations, continuing the tensions that had led to that war in the first place. No doubt a reaction to the scientific view of nature played some role in the romantic idealism of fascist movements.[18] According to Freud, the so-called death instinct (a psychological factor potentially leading to war) is a rebellion against the very strictures of civilized life. More recently, "postmodernism" is a kind of romantic revolt against the rationality of modernism and even against science, which has given us many ways to destroy ourselves but hardly taught us how to avoid destruction.[19] (In fairness, neither does any other cultural expression offer such instruction. No major religion, for example, has unequivocally opposed violence and exploitation.) Anti-scientism and other forms of romantic rebellion against the rule of reason have led not to balance but to war and fascism. Yet the challenge remains to find the balance that would obviate the need for the rebellion in the first place. The solution to the domination of reason is not unreason, emotionalism, mysticism, the absolute, or God, but something that is so foreign to the ideal of positive knowledge that it defies description and resists focus as a definable alternative. It is something that cannot be latched onto by reason and condemned as unreasonable. It cannot be made a thing at all. It is what I will call, in the concluding chapter, "the stance of unknowing."

Notes

1. van Fraassen, Bas C. 1999. "Structure: Its Shadow and Substance." PhilSci Archive. http://philsci-archive.pitt.edu/631/1/StructureBvF.pdf.
2. Lewontin, Richard. 2000. *The Triple Helix*. Cambridge, MA: Harvard University Press, p. 72–73.
3. Taleb, Nassim Nicholas. 2007. *The Black Swan*. New York: Random House, p. 69.
4. Cf. Smolin, Lee. 2013. "The Culture of Science Divided Against Itself." http://timereborn.com/wp/wp-content/uploads/2013/04/Brick_88_LeeSmolin1.pdf:
    "Literary intellectuals read and write texts and so engage their material with their fingertips, while artists, scientists, and engineers are *immersed in nature body and soul.*" (italics added) I would say rather that many scientists are immersed in theoretical ideas, in reading and writing texts; and experimentalists, like artists and engineers, work primarily with industrial materials!

5. McAllister, James W. 2011. "What Do Patterns in Empirical Data Tell Us about the Structure of the World?" *Synthese* 182:73–87.
6. Hartle, J. B. 2010 "The Quasiclassical Realms of this Quantum Universe." pp. 15–16. http://arxiv.org/pdf/0806.3776.pdf.
7. Barrow, John D. 1991. *Theories of Everything*. New York: Fawcett/Balantine, p. 156.
8. Homer-Dixon, Thomas. 2000. *The Ingenuity Gap*. New York: Alfred A. Knopf, pp. 6–7.
9. Taleb, *The Black Swan*, p. 227.
10. Brooks, Michael. 2012. *Free Radicals: The Secret Anarchy of Science*. New York: Overlook, p. 252.
11. Unzicker, Alexander, and Sheilla Jones. 2013. *Bankrupting Physics: How Today's Top Scientists Are Gambling Away Their Credibility*. New York: Palgrave Macmillan, pp. 99–100.
12. Smith, Tony. "Memes: Process Physics." http://meme.com.au/theoria/process_physics.html.
13. Tippler makes the point that we might expect to find integers or rational numbers at the base of a digital world, but we find instead numbers such as e and π. See Tipler, F. J. 2005. "The Structure of the World from Pure Numbers." Institute of Physics Publishing, *Reports on Progress in Physics* 68912.
14. Attran, Scott, and Joseph Henrich. 2010. "The Evolution of Religion: How Cognitive By-Products, Adaptive Learning Heuristics, Ritual Displays, and Group Competition Generate Deep Commitments to Prosocial Religions." *Biological Theory* 5:18–30, 22–23.
15. Pickering, Andrew. 1984. *Constructing Quarks: A Sociological History of Particle Physics*. Chicago: University of Chicago Press, p. 7ff.
16. Baggott. Jim. 2013. *Farewell to Reality: How Modern Physics Has Betrayed the Search for Scientific Truth*. New York: Pegasus, p. xii.
17. Within game theory, as the branch of mathematics devoted to economic management, the very definition of "rational" means the ruthless pursuit of self-advantage!
18. Prigogine, Ilya, and Isabelle Stengers. 1984. *Order Out of Chaos: Man's New Dialogue with Nature*. New York: Bantam, p. 6.
19. Václav Havel, quoted in Holton, Gerald. 1996. *Einstein, History, and other Passions*. Boston: Addison-Wesley, p. 33.

Part Three

Maker's Knowledge

11

The Book of Nature

> *"In the beginning was the Word."*
> —John 1:1

> *"There is nothing outside the text."*
> —Jacques Derrida

The Greek Heritage and the Logos

Preliterate goddess religions had revered nature itself, not an abstract principle *behind* nature. According to the Bible, in contrast, a God separate from nature had "spoken" the world into being. Divine creation is an act of fiat, invoked by the commanding word—as in "Let there be light." This Judaic concept merges with the Greek concept of Logos, as the rational principle.[1] The Greeks called any reasoned text Logos, but generalized the concept to mean reasoned thought itself. Early Christians interpreted it as the *divine word* (God's reasoned text)—which was not necessarily assumed transparent to human thought.[2] There were thus two records testifying to the Creation: the Bible and the natural world itself.

Both the Greek notion of Logos and the Jewish emphasis on scripture provided a powerful basis for a metaphorical understanding of nature as textual exegesis, which came to dominate medieval Christianity.[3] This understanding of the Bible as both written history and covenant dovetailed with medieval fatalism. "It is written" meant "it is destined." The fixed content of a text became the early template of a deterministic philosophy. While speech flows irrevocably in one direction, pacing time, one can search back and forth within a text at will. While speech is ephemeral, text is reversible and outside time.

God authored both the world and scripture, which stand therefore in a special relationship as twin sacred texts. In Christian Europe, the natural world was referred to as the "Book of Nature," which was considered to complement the Bible as a guide to divine will.[4] Holy writings and nature itself were alternative expressions of God's message and purpose for humanity.

Science inherited from the Greeks the idea of nature as deductive system, on the model of geometry. From the biblical tradition it inherited the parallel idea of nature as text. Each of these complementary notions reflects a belief that the world is the result of an intentional creative act. Together they would affect the treatment of nature in science and by society for generations to come. Early science hardly breaks with medieval tradition concerning the Book of Nature; quite the contrary, it implicitly regards nature as a text to be deciphered.[5]

Greek philosophy had embodied a faith that the overwhelming reality of nature could be trumped by intellectual means. By embracing a version of determinism, formalized as logical necessity, the ancients tried to get inside the apparent fatalism of the world, to understand and be able to predict its workings. Their very concept of knowledge served to bridge the apparent gulf between subject and object, self and world.[6] This bridge is built from the human side: in knowing, the subject actively assimilates the object in the form of a representation, image, or internal map. This parallels the orthodox theism of the Semitic religions, which assimilate divine mystery to a text. Yet its intent was quite different: freedom from the supernatural. It stands in contrast to dissolution of the self in some mystical traditions—where the self is assimilated in the object, instead of the other way around.

Veneration of the Written Word

Writing fosters awareness of the structure of language as a medium, in contrast to its meaning. It allows verbal expressions to be approached statically and objectively, as timeless objects viewed externally rather than as the moment-to-moment unfolding of meaning in speech.

A striking aspect of the medieval mind is its awe for the authority of written texts.[7] Though many scientific and philosophical works from the classical period had been lost in the Dark Ages, legal traditions from antiquity were preserved in writing straight through to modern times.[8] Logic developed within the context of law, which supplied ample practical use of conditionals and modalities. Law itself loosely comprises a deductive system, with decisions derived from precedent in much the way that theorems are derived from axioms or corollaries from theorems. While embodying capacities for reflection, and providing a model for deductive thinking, law is also bound to practical decision and remains sensitive to facts, the weighing of evidence. Many of the early scientists and other Renaissance intellectuals had been involved with law or medicine, or both. (At the time, these were less specialized than now and formed part of what we now call a liberal arts education.) Both were domains of expertise in which professional opinion is valued and certainty is achieved through an established body of principles and practices. Both had their theoretical and applied sides, and both contributed directly to the development of science.

As well as religious writings, classical works continued to be venerated, not least because knowledge in antiquity was supposed to be closer to the original knowledge of Adam before the fall, less subject to the corrupting influence of time—a major consideration when manuscripts had to be repeatedly copied by hand.[9] This was not simply a matter of respect for the past, but also reflected the certitude that attends a closed system of meaning, embodied in texts. Above all, it involved a symbolic way of thinking, and an attitude toward life and nature very different from the modern one. Prior to the spread of printing, inconsistencies between different manuscript copies of texts had led to the scholastic method, which sought to resolve contradictions among compared texts by finding a higher ground of general principles that could serve as a basis for textual interpretation. This led away from literal readings to the sort of allegorical speculation for which medieval scholasticism is noted.[10]

Medieval Christians thought allegorically and interpreted texts freely, imaginatively, and above all symbolically. They mined scripture for spiritual lessons they could apply in daily life, and brought the same approach to the Book of Nature. Even the geographical maps they produced were not literal, in a modern sense of faithfully representing spatial relationships and topological features; rather, they served symbolic, decorative, and didactic purposes. Hence, for example, the three continents of the old world were associated with the Trinity.[11] The natural world was treated as a parable whose meaning was spiritual, and whose details could be examined for clues to the mind and intentions of its Creator. This reflected the biblical view of man as the centerpiece of Creation, which was tailor-made to provide moral instruction as well as sustenance.

The context for reading the Book of Nature was the eschatology of medieval Christianity. Study of nature meant attention to the miraculous and portentous; the Book of Nature, like scripture, was read for its prophetic value, not out of dispassionate curiosity. People were more interested in locating their generation in the biblical timetable, counting down toward apocalypse, than in the physics or cosmology of a world that was not destined to endure. The printing press contributed to widespread interest in prophecies and biblical interpretations of natural portents like earthquakes and comets.[12]

For most of human history, people had wondered about the significance of natural events for the present or future. Whether by astrology or divination, priestly specialists undertook to read natural signs for an appropriate course of action. The Christian allegorical tradition was no exception; it was a system of symbolic meanings for interpreting nature and scripture in terms of human consequence. A personal relationship to the Creator implied that natural phenomena were intended to teach moral lessons; hence, their allegorical status.[13] Medieval understanding of disease mixed causal and intentional factors, which were often at odds. The view of disease as a divine visitation did not meld easily with the naturalistic ideas of both Greek and Roman medicine. Hence, medieval Christians typically believed that events could be simultaneously natural and divine and that God used the workings of nature to achieve divine purposes.[14] Allegorical thought involved an integration of diverse concerns. The idea of deciphering nature as a system in its own right, bearing intrinsic interest independent of implication for human spiritual destiny, would represent a major shift toward a more literal and less allegorical understanding.

Protestant movements brought with them such a shift in attitude. Scripture was interpreted more literally, less symbolically. Like Islam, Protestantism tended to reject the image, associated with Catholic icons, in favor of the written word. Protestant natural philosophers rejected the priesthood and hierarchy of the Church, and with it slavish devotion to past scholarship. They wanted to read and interpret the Bible for themselves—and equally the Book of Nature. They sought a fresh start, a break from the convoluted metaphysics of the Church intelligentsia, whose academic discussions on unverifiable theses seemed to have little bearing on ordinary life. They rejected the right of the clergy to interpret scripture on their behalf and sought direct personal access to its meanings. Printing brought Bibles within reach of the masses. In a period in which people widely claimed the right to interpret (and could pay dearly for a "wrong" interpretation), it now seemed safer to focus on obvious literal meanings.

Religious wars and the Counter-Reformation provided good reason to shy away from the freer interpretations that had characterized an earlier mentality. The Royal Society, for example, though initially including many radical social thinkers, soon settled down to a conservative program. Its members were fed up with war, disorder, and unrest. These were mostly Anglican monarchists who viewed God's relationship to nature through the metaphor of a sovereign lord over his domain. They were keenly interested in order, in both the natural and human worlds, which were supposed to be integral as twin expressions of divine will. Protestant emphasis on literal interpretation spread to other areas of culture in Restoration England. There was a general rejection of "fancies" that did not correspond to literal reality, whether in literature, poetry, drama, or philosophy. The same attitude was applied in science, which became the standard of unadorned, impersonal, mathematically precise description.[15]

Nature as Text

The vision of the world as *text* is closely related to that of the world as a divine *artifact*—indeed, as a machine. Both are made by design. All artifacts, including texts, possess primarily the reality assigned by their creators and users. Like other artifacts, including machines, a text is a finite, self-enclosed intentional product of definition. It contains no more than was explicitly inscribed by its author along with deductions implicit within that. If *nature* is a text, then it should be as predictable as a machine, as searchable as other texts, and subject to the methods of textual interpretation that were applied to scripture.

Nature is implicitly approached as a text in Greek deductive thought. It is explicitly both artifact and text in Christian thought, being created or authored by God. Medieval intellectual life was a culture of the text, in which scholarship typically invoked reference to other textual commentary. Not only did the individual manuscript define a finite system in its own right but the corpus of texts collectively constituted a larger—and closed—system of thought. This provided a basis on which to approach the natural world with confidence, perhaps temerity. Thus, for example, in place of the careful observation Aristotle had brought to the study of nature, medieval scholars would merely consult his texts and others' commentaries on them.

Textual exegesis, as understood in medieval times, revolved around spiritual life. With the possibility of multiple and symbolic meanings, it was a free interplay of the document and the interpreter, intended to guide daily living. The literalism of the Reformation replaced this allegorical approach with an attention to the document itself, as though it were fixed, without reference either to other documents or to context, independent of interpreters, and bearing only a singular meaning. Natural philosophers adapted this approach to the Book of Nature. Thus, in part to distance himself from political repercussions of his own writings, Galileo craftily extols nature as a book that can be read "directly," for which interpretation is irrelevant. He proposes to replace the allegorical *why* of religion with the *how* of mathematical description. Galileo was the first to introduce the terse style that would thereafter characterize scientific texts—thus expressing a watershed between the medieval and modern mentality.[16]

Just as the Bible was to be taken more literally in the Reformation, so was the Book of Nature. Both were to be freed from the speculative excesses of medieval scholasticism.[17] The textualism characteristic of medieval thought had led to the practice of referring primarily to texts for information about nature, rather than to nature itself. This fostered a proliferation of absurd misconceptions about the appearance of wild or exotic creatures. But this practice eventually gave way to a newfound respect for facts that could only be ascertained by actual observation—by directly reading the Book of Nature for oneself. On the other hand, by the beginning of the seventeenth century, some Protestants had begun to regard the Bible as an infallible source of knowledge about nature.[18] And, conversely, the Book of Nature continued to be consulted, like the book of Revelations, for its prophetic significance.[19] Newton and some of his contemporaries recognized no essential difference between their scientific and textual studies.

Textualism

Textualism, as I am calling it here, is the notion that nature itself is in some sense literally a text—the equivalent of, indeed a version of, deductionism. This is a persistent and largely unacknowledged theme inherited from the medieval concept of the Book of Nature. Medieval thought held that the mind and will of God could be understood through his dual creative expressions: scripture and nature. Divine law was given to man directly in scripture (to regulate human affairs) and indirectly in the Book of Nature (to regulate the physical creation).

The written word—and particularly the *printed* word—could stand on its own, out of context and detached from the speaker or author. When applied to the divine word, this alienation reflected the broken covenant, the estrangement of God after the fall. Despite the hope of spiritual restoration through worldly means, the Renaissance Christian faced a universe that had to be decrypted like an intercepted enemy message.[20] The colorful medieval style of textual interpretation, which had sought out and compared crucial symbolic meanings in copied manuscripts, gave way to a more consolidated dogma, reflecting the standardized ubiquity of the printed text.

Christianity in general had rejected the Greek tradition of intrepid speculation, which had been attended, nevertheless, by responsibility for one's views and a moral association of words with the character of the speaker. In the Christian tradition, it was *God* who spoke through his interpreters, rendering mere human opinion irrelevant when not blasphemous. A major difference between speech and written language, moreover, is that a text is a *thing*. It is present all at once, of a piece, autonomous and independent of the speaker. Speech, on the other hand, is necessarily presented sequentially by the person speaking. A text was originally a record or reconstruction of speech. While it is normally read in order, it need not be considered a linear sequence at all, but may be deconstructed, passage by passage, manipulated or taken out of context. As an abstraction, it exists outside time. In this respect it resembles the closed reversible system of physics. Both narration and causality make time flow in a single direction.[21] (In fact, causality *is* a narration—a story about the relationship of things in time.) A text as a free-standing entity, however, can be read out of order. It may be considered a closed logical system, without external reference. It may be inspected for internal structure, pattern, or logical relation, apart from flow, direction, context, or meaning. Ultimately, it may be considered pure "information"—a mere collection of zeros and ones.

Text as Formalism and Formula

Whatever else it is, science is a textual narration. We have already noted that determinism characterizes deductive systems, not natural ones. The historical context for this is the fact that biblical prophecy is a function of the Bible as a text. Prophecy requires that one can move freely backward and forward in time, *as within a text*, because all relations within the entire content are already fixed—in history as in text. For actual historical time was identified with biblical narrative as a historical record. The ability to search the text is conflated with the ability to search actual time. Events are predictable in a deterministic system for similar reasons, because it is a deductive system, a product of definition like a text. Whether in the mind of God, in the mind of Shakespeare, or in the mind of the physicist, texts—while open to interpretation—are determinate and searchable in a way that nature is not.

The ideal of objectivity is to occupy a place outside nature and time, to be above it as the mind of God is above his creation—as an author (or reader) is in relation to a text. As God became divorced from his creation in Deist thought, so the scientist assumed the role of a disembodied consciousness outside the natural systems studied. Thus, the view of the universe as a self-contained machine, running without the ongoing support of divine will, represented a new textualism. In that light, the Book of Nature could be studied to advantage in syntactic terms, specifically in the language of mathematics, the most efficient way to express and interpret the apparent syntax of the world. Mathematics and a standardized scientific method were able to unify knowledge, but at the cost of divorcing it from spiritual concerns—indeed, from semantic meaning. As God removed from nature, so did man.

Whether the text is written by God or by the theoretical physicist, the advantage of presuming nature itself to *be* a text is the prospect that it can be exhaustively formalized.[22] Hence, it is predictable, literally by formula. "Information" is the modern constituent of "text," and the world is held to have a definite information content, like a text. Computer programs, algorithms, and equations literally *are* texts. The belief that nature consists fundamentally of information or computation is the current version of an ancient obsession.

Though modern science is not so different from medieval textualism, the mechanist worldview did not lend itself to traditional medieval analysis, in terms of the specific "nature" inherent in each thing or kind. Aristotle had deliberately made little distinction between natural and artificial, as far as teleology goes; but his examples apply more to biology than to physics, as the new scientists had begun to conceive it—on the model of clockwork. This new model had a place for teleology quite different from Aristotle's: If nature was an artifact, then it must embody and reveal the intentions of its designer. At the same time, if nature was a mechanism, it was subject to detailed analysis of the efficient causes between parts, which could be considered apart from the purposes of its creator.

Notes

1. Barbour, Ian G. 1990. *Religion and Science: Historical and Contemporary Issues*. New York: Harper, p. 241. The first major Christian apologist, Justin (died ca. AD 165) held Christ to be the divine Logos, identified with the power

of reason. See Lindberg, David C. 1986. "Science and the Early Church." In *God and Nature*, edited by Lindberg and Ronald L. Numbers. Chicago, University of Chicago Press, p. 23.
2. Freeman, Charles. 2005. *The Closing of the Western Mind: The Rise of Faith and the Fall of Reason.* New York: Vintage, p. 23.
3. Cf. Bono, James J. 1995. *The Word of God and the Languages of Man: Interpreting Nature in Early Modern Science. Vol 1: Ficino to Descartes.* Madison: University of Wisconsin Press, p. 11.
4. Menuge, Angus J. L. 2003. "Interpreting the Book of Nature." *Perspectives on Christian Faith* 55:88–98.
5. Bono, *The Word of God and the Languages of Man*, p. 5. On the other hand, for Margaret Cavendish, Duchess of Newcastle, nature was not God's text so much as an author in her own right.
6. Dewart, Leslie. 1969. *The Foundations of Belief.* New York: Herder & Herder, p. 71.
7. Barrow, John D., and Frank J. Tipler. 1986. *The Anthropic Cosmological Principle.* Oxford: Oxford University Press, p. 46.
8. Franklin, James. 2001. *The Science of Conjecture: Evidence and Probability before Pascal.* Baltimore: John Hopkins University Press, p. 4.
9. Menuge, "Interpreting the Book of Nature," p. 90.
10. Franklin, *The Science of Conjecture.* p. 17.
11. See the McGill University project, "Making Publics" or the CBC Radio *Ideas* series episode based upon it, "Origins of the Modern Public." http://www.cbc.ca/player/Radio/Ideas/Full+Episodes/2010/ID/1528896270/.
12. Webster, Charles. 1982. *From Paracelsus to Newton: Magic and the Making of Modern Science.* Cambridge, UK: Cambridge University Press, pp. 16–17.
13. Barbour, *Religion and Science*, p. 6.
14. Lindberg, David C. 1992. *The Beginnings of Western Science: The European Scientific Tradition in Philosophical, Religious, and Institutional Context, 600 BC to AD 1450.* Chicago: University of Chicago Press, pp. 320–21. Hence, the Church fathers insisted on the divine origin of secular medicine.
15. Merton, Robert K. 1938. "Science, Technology, and Society in Seventeenth Century England." *Osiris* 4:378.
16. Koestler. Arthur. 1960. *The Watershed.* Lanham, MD: University Press of America, p. 186.
17. Merton, "Science, Technology, and Society in Seventeenth Century England," p. 429.
18. Barbour, *Religion and Science*, p. 13.
19. Webster, *From Paracelsus to Newton*, p. 10.
20. Bono, *The Word of God and the Languages of Man*, p. 84.
21. Taleb, Nassim Nicholas. 2007. *The Black Swan.* New York: Random House, p. 70.
22. On the other hand, this may be one motivation for postmodern deconstructionism, which insists that reality should be interpreted as though it were a text: to restore the latitude one has in interpretation that cannot be supposed in a realist view of nature.

12

The Religious Origins of Science

"Science began as an outgrowth of theology, and all scientists, whether atheists or theists ... accept an essentially theological worldview."
—Paul Davies[1]

"What is man that he is mindful of the universe?"
—John D. Barrow

The Rise of Reason

Western secular culture is more or less a product of the Reformation. This was a time of great change and uncertainty in which beliefs were widely questioned and could be fatal. Trade increased contact between diverse cultures, creating a new urban middle class and a capitalist economy. Voyages of discovery found whole new continents with new riches, unfamiliar plants and creatures, and strange inhabitants. The discoveries of Galileo shattered the celestial spheres and the separation of above from below that had characterized a millennium. All of this contributed to a renewed interest in the natural world. It was also the time of the rise of literacy and printing so that people were widely confronted with a multitude of perspectives and values, including the reasoned ideas of the ancients. It was a time, therefore, in which the received knowledge of the walled medieval world could no longer be relied upon. Faith alone was no longer adequate; people were forced to think for themselves, and began to claim the right to do so.

The tradition of reasoned argument was inherited from the Greeks, for whom it had been an antidote to superstition. Far from expressing animism, Thales's "gods in all things" became Aristotle's "natures," which resided inherently in things as their own inner necessity, knowledge of which could yield some authority over them.[2] The Greeks recognized that experience depends on both object and subject. This insight promised some control over one's personal experience, thought processes, feelings, and the inner workings of the psyche, inspiring the organization of thought in disciplined ways, imposing its categories upon the natural world. Yet Greek belief in the reality of nature still entailed a fatalistic power over human and even divine affairs. Christianity took exception to a view that implied no free will for either man or God. The Christian concept of nature, as specially created for human benefit, overthrew this notion, eventually in favor of a technological science based on experimental intervention, yielding knowledge and power over nature. Thus, Christianity opposed the autonomous reality of nature in order to uphold divine and human freedom and authority.

Through contact with the Arab world Christianity recovered the lost teachings of Greece, preserved in Arabic translations. It was through these translations, owing to the difference in language and mentality, that the notion of *contingency* entered Christianity, contrary to Greek fatalism.[3] For the Arabs, the very existence of finite things entailed their ability to be encountered accidentally. In contrast to nature, God alone is necessary and real by virtue of himself. For Newton too, at the beginning of the Age of Reason, God would be the only self-subsisting reality.

Despite their competition in the early modern period, faith and reason have a common basis in accepted premises. However, the premises of faith were embraced through tradition, scholarship, textualism, historical accident, or political imposition; those of reason were grounded in direct experience, intuition, and common sense. Faith had to be embraced whether or not it made sense to the individual, and despite the fact it might contradict other creeds. Reason, supposedly universal in principle, could make sense to all and provide a basis for consensus.

The very fact that the Christian view of nature gave way to the secular scientific view intimates a continuity of intent, method, and ethos between them. There is but a fine line between the faith-based biblical dominion appointed to man over Creation and the reason-based domination of nature through technology; between the quest for godliness and the quest for god-like powers. The gradual transfer of creative authority from God to man is intimated on the ceiling of the Sistine Chapel, where Adam appears as nearly the equal of God, and it is unclear which has reached out to bring life to the other.[4]

Thus, the emergence of modern science was more or less continuous with its religious roots. Its early practitioners were devout men. It was sanctioned by Puritan theologians especially. It not only sprang from religious soil, but took on some of the characteristics and goals of theology. It became a means of spiritual inquiry, an alternative not only to scholasticism but to scripture itself. It was an alternative form of revelation. While medieval Christianity had devalued nature and its study as pointless or even sinful, the post-Reformation attitude saw in the material world signs of divine intention that could and should be studied as a religious duty.

Augustine had conceded that knowledge of nature was useful to elucidate Christian doctrine: science was the handmaiden of religion, since God was the cause of all things. The Enlightenment philosophers had a far more detailed notion of cause in mind. Their new "arm's length" conception of God's relationship to nature had the moral advantage of alleviating divine responsibility for embarrassing aspects of the Creation, such as noxious creatures, disease, and natural disaster. They tended also to intellectualize faith, shifting emphasis from personal salvation to a more detached understanding of the role of the creator God. This opened the way for secular salvation, through social institutions and technology.

Part of the motivation for the philosophy of mechanism lay in rejecting medieval notions that confounded the natural with the supernatural—for example, in fantastic creatures and various concepts of natural magic. Mechanizing matter served to separate it more clearly from the spiritual realm.[5] Reason, combined with careful observation, could hinder religious faith but was suited to deal with the material world. In his 1758 treatise on dynamics, d'Alembert shifted the ground from theology to theory by suggesting that nothing could be proven from the study of nature, one way or the other, about the existence or freedom of God. The germane question, he thought, was whether the laws of nature are empirical or derivable from first principles. Both d'Alembert and Hume proposed that scientific knowledge could progress without metaphysical certainty. That is, one could know mathematical relations between things without knowing their ultimate causes or natures.[6]

From Aristotle to Deism

By dismissing the power of supernatural agencies, some lines of Greek thought focused on the immanent reality of natural things, which contained their own internal powers and were the source of their own being. This classical inheritance was overturned by Christianity, which favored a Platonic line of thought, depriving found things of their inherent natures and leaving them with only the reality conferred on them by their supernatural Creator.

The medieval concept of the natural world had been strongly influenced by Aristotle, for whom science was the study of the individual "natures" of things. These were powers within natural things themselves, constituting the source of change. In contrast, artificial things possessed no such inner power, being the mere result of external agency. In rejecting Aristotle, Renaissance science retained only the "efficient" causes that operated *between* things; these were held implicitly to be the sort of change initiated originally by an external agent, ultimately the first cause. This understanding of causality favored a mechanist view of nature and the rise of machines in society, whose successful use in turn reinforced the mechanist view. The type of cause arising from *within* things themselves was absent in the mechanist cosmos, having been transferred to the divine will operating from outside nature.

Medieval theologians had been anxious to preserve divine freedom of will—as expressed, for example, through miracles. Miraculous appearances and transformations, however, posed questions about what would later be called the conservation of matter. The transubstantiation of the Communion host, for example, raised the question of *what* exactly is transformed, given that the sensible properties of bread and wine remain unchanged. To assume that appearances are totally independent of any real substratum (i.e., that God can make it appear to us however he wishes) was repugnant to a rationalist sensibility. The issue came to a head in the sixteenth century and was one of the controversies that brought about the Reformation, which "picked apart heaven and earth."[7] The Protestant position was that transubstantiation must take place within the subject—the worshipper—not externally in the object as formerly held. What was important was the faith of the believer, who must directly experience and be persuaded of the divine presence in the course of the sacrament.

Personal experience and opinion—subjectivity—came to matter in a milieu where doctrine could be publicly disputed. What we now call public opinion had relevance for the first time and was needed to endorse the political order.[8] (The pulpit was persuasively used toward that end.) People now had to decide for themselves questions the Church had formerly decided for them. This led to widespread questioning of how institutions, doctrines, and knowledge generally could be justified and to the general possibility of skepticism. One needed not only to know but also to know *how* one knows.

Miracles went contrary to the intent of classical philosophy, which was to show a rule-bound world regulated according to reason and free from superstition. The conflict between natural causality and miracles had weighed upon medieval thinkers and produced a spectrum of compromise solutions. The consensus was that God chose to rule the world through created laws, but nevertheless retained the power to suspend them at will.

For Aristotle, substance and form were simply dimensions of the being of things. Western philosophy would later say that aspects of form are imposed by the human mind. But for the early moderns, form was clearly imposed by the mind of God. Matter needed no internal principle of change or self-organization. Once created and set in motion, it could be left on its own, though it might wear out or wind down like a machine and need to be restored periodically through divine maintenance. The guiding principles that forced passive matter to behave in accordance with its decreed "form" were the laws of nature, which governed matter by divine authority in much the way that human laws govern the affairs of men. On this understanding, it was spiritually as well as materially beneficial to investigate natural phenomena as manifestations of divine will.

In keeping with such ideas, the new scientists considered teleology to be imposed on nature rather than immanent within it; nature itself was no longer thought to be striving toward any end of its own.[9] The idea of God as final cause—the supreme good toward which Creation aims—was gradually displaced by the idea of God as first cause, the hand that initiates a chain of efficient causes in a domino effect. The Aristotelian idea that individual natural things possess their own innate reasons for being gave way to the idea that nature as a whole was a mechanical artifact, designed and set in motion by the provident beneficence of the Lord.[10]

The early Church preferred Plato to Aristotle and knew little of the writings of the Greek materialists. In early medieval Europe, the goal of nature study was to glorify God. This was more compatible with Plato's philosophy than with Aristotle's, which taught a world without beginning or end, a perishable soul, and inherent properties of matter independent of divine intent. While Greek thought in some ways implied the immanent reality of nature, in others it circumscribed natural reality as logically necessary and containable in deductive systems. Plato, in particular, had also advocated a monotheistic God and upheld the immortality of the soul. Yet some theologians denounced the entire classical heritage as pagan.[11] The struggle between faith and reason continued throughout the fourteenth century, often centering on the issue of God's free will and absolute power, and tending to erode confidence in mere human powers of thought.[12]

Neo-Platonism downgraded matter as inactive, in contrast to spirit or mind, which was the active principle—particularly the mind of God, the first cause. This dualism was the source of early confusions concerning the concepts of inertia, mass, and force.[13] The Deist tradition founded on Neo-Platonism appealed to Enlightenment philosophers, for whom natural causation could be described in terms of laws, while the role of the Prime Mover was reduced to supplying initial conditions.

Though space may have been the venue of physical reality, and Newton's "sensorium" of God, the fact that God permeated space did not mean that he was to be identified with nature. Thus, Newton hastens to add that "this Being governs all things, not as the soul of the world, but as lord over all." Boyle expressed similar sentiments, insisting on a radical distinction between creator and creation, and that powers accredited to nature only detracted from the glory of God.[14] Perhaps, as Margaret Cavendish suspected, they also detracted from the glory of the Enlightenment thinkers.[15] While a mind operating the system of nature from outside was no embarrassment to Newton or Boyle, it eventually became so to others who elaborated their thought. Thus, Laplace had no need for "that hypothesis." Like the ancient Greeks, scientists now wanted to know how nature works without reference to the supernatural. The nineteenth-century aether took the place of the sensorium of God. Yet the immanent reality of nature continued to be ignored, and the problem of accounting for initial conditions—how it all began—remained unsolved.

Patriarchy and Monotheism

The Great Goddess had been nature itself, not an abstract principle or external creator. There can be little doubt that the rise of monotheism coincided with the rise of patriarchy, and hence with a masculine creator god *behind* nature and separate from it. This implied a more abstract concept of nature along with greater psychological distance. It seems unlikely that the ancient goddess cult could have been a monotheist religion of transcendence (with a female god instead of male); a transcendent earth spirit would be a contradiction in terms.

The Greeks, who were polytheistic, democratic, and somewhat anarchistic, lacked the concept of a unified external lawgiver to dictate the behavior of things. As in many other cultures, the various gods of the Greek pantheon represented competing aspects of chaotic nature and human psychology, while only loosely subject to the central authority of Zeus. Their gods appear comic to us in their obvious humanity and may not have been taken seriously by the Greeks themselves. Plato held reason—an abstract principle—to be the force responsible for an orderly cosmos. While he had emphasized the *idea* as the primary "reality," from which the created thing is merely an imperfect copy, Christian thinkers embraced Platonism for their own purposes, to emphasize the creative freedom of the divine Craftsman. Both denigrated alike the reality of nature (the Craftsman's medium) and the world of appearances. The Ideas needed a divine mind to think them, completing their projection outside the human psyche. The Judaic belief in a single willful creator complemented the Greek expectation of a rationally comprehensible universe, which was a prerequisite for deductive science. Christianity added the touch that problems of induction could be solved ultimately by acts of faith.[16]

Christianity melded the Greek and Judaic traditions, through the filter of Arab scholarship. The scriptures represented a covenant and a history, unfolding as a grand human story against the backdrop of an irreversible linear time. This contrasted utterly with the cyclical time of preliterate cultures. Classical physics inherits the contradiction between these views—enshrining the individual's localized historical point of view alongside the idea of reversible time.

According to the ancients, natural bodies contained within themselves certain properties and tendencies to move in certain ways. Similarly, medieval and early Renaissance alchemists believed in powers residing within matter, sometimes defending pantheistic views in which God was identified with nature.[17] Newton and his contemporaries rejected such views in favor of mutual influences between otherwise inert things, in a chain of causes that had to be initiated ultimately by a masculine Power residing outside it. Descartes regarded the mathematical truths of physics as divinely revealed—a Platonic belief that lives on in our age.[18]

The European Context

Religion favored the growth of science in Europe for several reasons. First, the idea of laws of nature required an external lawgiver, which the Judaic tradition provided. A god separate from nature could create general laws and yet retain the right to specify details (and even to break the laws). This meant that nature was contingent rather than logically necessary (as the Greeks had believed), so that its details could only be discovered through observation. Moreover, if God resembled man, then the divine Creation could be rationally understood. This gave hope for practical knowledge of a negotiable world with knowable properties and rules. With a beginning, middle, and end—like scripture—nature could be read as a text. Printing arose in Europe, which made actual texts widely available. Christian dogma assimilated the cosmos to metaphors within the human sphere, in contrast to an eternal cycle of repetitions or an inscrutable mystery at the mercy of chance. It unified nature as the creation of a single will, which could be addressed through a personal relationship, on the one hand, and scientific protocol on the other.

The Renaissance thinker was able to integrate knowledge from every source under the divine aegis: reason, observation, experiment, tradition, and scripture. All sources would converge in perfect knowledge, reflecting God himself.[19] The fact that created things, once created, have a certain autonomy removed them a degree from divine sanction, placing them more at human disposal. The implication of St. Thomas's philosophy was that humans too are creators, with some control over their destiny and dominion over the natural world.

In the antique world, learning had been segregated into rival schools of thought. Christianity exerted a unifying influence, so that a consistent curriculum was taught in medieval universities.[20] Such pedagogic unification no doubt helped to facilitate an interest in objective truth, so that emphasis could shift from metaphysical debate to something more resembling the modern notion of reasoned theory tested by observation. On the other hand, the collapse of central Church authority during the Reformation meant that people had to sift arguments for themselves. It meant that religious, political, and even scientific questions were open to debate, subject to evidence. Any given position now required the support of others, making it a public matter and no longer simply a matter of private belief.

The Rejection of Scholasticism

Medieval thinkers had generally been interested in speculation and textual exegesis more than in experimentally testing hypotheses.[21] Theirs was a self-enclosed system of thought reflecting the shutting down of intellectual inquiry during the so-called Dark Ages.[22] This, in turn, had reflected the suppression of thorny conflicts among contending creeds, in favor of a Christianity unified under a politically motivated orthodoxy. But the reconciliation of contingency and divine potency in the early modern period was favorable to experimental science: nature could only be known by observing it carefully. Moreover, perceiving the order meant isolating causes in artificially contrived situations—experiments—which alone could reveal mathematically expressible relationships.

The Scientific Revolution was bound up intimately with the Reformation, which proposed to democratize knowledge of nature, as of scripture. This was approached in several ways: by addressing the commons of nature; by equalizing observers in principle through a standard experimental protocol; by sharing knowledge freely. The Protestant scientists rejected the priesthood and hierarchy of the Church and the slavish devotion to past scholarship. Both the Bible and the Book of Nature were to be taken more literally and freed from the fancies of medieval interpretation. Self-contained metaphysical systems were to take a back seat to hard facts.[23] The new scientists wanted a fresh break from the convoluted and ingrown scholasticism of an elite whose endless discussions on unverifiable theses seemed to have little bearing on ordinary life. In addition, scientific method was appealing because experiments could be reproduced. Repeatable experiments also implied that natural phenomena could be controlled, even re-created at will, confirming a semi-divine status for human beings.[24]

Faith Rationalized

Medieval intellectuals sought to demonstrate the compatibility of reason and faith. Roger Bacon advocated science not just for its humanitarian benefits, but also for its ability to support religious dogma, as the handmaiden of religion. Like Thomas Aquinas, he dismissed conflicts between reasoned argument and Christian dogma as matters of faulty translation or interpretation of texts, insisting that natural philosophy is God given, so that there can be no true conflict between reason and faith.[25] Descartes would echo this sentiment by insisting that God would not permit us to be systematically deceived by our senses. Yet a general unease in regard to sensation and reason alike underlay philosophical argument: for only revealed truth, as transmitted by the Church, was considered absolutely certain, guaranteed by God. Throughout the early modern period, conflict with Church doctrine was circumvented through the device of considering contentious scientific ideas to be merely hypothetical, not concerning fact. While diplomatically motivated, this convention set the stage for the modern concept of the scientific model: a hypothetical device obliged to accord not with church doctrine but with experiment. Sincerity apart, the religious arguments of early scientists lent credibility to the new sciences and helped to gain their acceptance.[26]

Augustine had taught that faith must precede reason where reason cannot yet grasp the truth. The role of faith is to "purify the heart and make it fit to receive and endure the great light of reason."[27] Conflicts between reason and dogma came to a head in late thirteenth century, with condemnations of Aristotle. These focused especially on points that had to do with divine and human freedom, in contrast to the stubborn autonomy of nature. The eternity of the world was denied in favor of the soul's immortality. The "world soul" was denied in favor of individual personal souls and the personal creative agency of an eternal God outside nature. Logic was denigrated in favor of free will, human or divine. While the ancients had attempted to describe the world as it *must* be on logical grounds, their counterparts in the Church insisted that the world is whatever the Lord wills it to be, even moment to moment.[28]

The spiritualism and unworldly concerns of the medieval period had not been generally conducive to science as the study of this world. Indeed, this world continued to be considered evil by Calvinists and Catholics alike. While the latter prescribed withdrawal into monastic life, however, the Protestant solution was to remake the world itself and one's relation to it through relentless hard work.[29] Human activity should glorify God, and an important way to do so was through labors that would benefit society. Good works were also pursued as an outward sign of inner grace. The need for diligence in such pursuits promoted a careful and systematic approach, which also avoided the temptations of idle hands.[30] These attitudes and qualities fostered a hands-on approach to nature.

The Religious Platform of Early Science

Ostensible religious devotion was a price paid more or less willingly by most early scientists, whether Catholic or Protestant, in exchange for the needed sanction by religious authority for their scientific pursuits. Like Galileo, Francis Bacon had called for a "direct" approach to nature, consistent with the Protestant approach to biblical study.

He had a particular method in mind through which to avoid the pitfalls of knowledge passively derived from the senses: the scientific method of careful experimentation. The creative role of imagination is to be grounded in cleverly designed interventions to "interrogate" nature. Knowledge of nature is thereby redefined as answers actively pried from her through experiment.

Bacon's grand program was both religious and humanist: to restore mankind to its proper station prior to the fall. Society could do this, he believed, by pursuing the biblical dominion over nature. Since God is the power behind Creation, it is ultimately through imitating this creative power that mankind may recover its rightful status. The transcendent being of God, separate from the world in the way that mind is separate from body, meant that nature could be studied, manipulated, and freely exploited; it was not to be revered as immanently divine. Adam's original state of innocence, which supposedly included perfect knowledge of nature, could now be recovered through science and technology. Mankind's separation from nature—resulting from sin and the expulsion from the Garden—could be turned to advantage. It mirrored the divine separation from nature, empowering human domination and thereby (if paradoxically) promising the restoration of innocence. Thus knowledge and power, rather than moral virtue, became the new basis for human salvation.[31] The creator God gradually eclipsed the redeemer God as the focus of interest, as reason displaced faith as a path to knowledge of both God and his Creation.[32] In the hands of the seventeenth-century rationalists, religion became more of an intellectual exercise than a living experience of faith. In contrast to the fatalism of the medieval period, the modern idea of *progress* was now afoot in a new humanism that placed faith in technology, the future, and humanity's ability to secure its own way toward knowledge and salvation through worldly effort.

Even throughout the nineteenth century, religion continued to be a major force in society and central in the thought of many scientists. Hence, Darwin's personal struggle regarding his biological ideas and his traditional faith. Lord Kelvin (William Thomson) saw in the first law of thermodynamics (conservation of energy) evidence of the permanence of the Creator, as opposed to the second law (increase of entropy), which testified to the transience of the Creation.[33] Some saw in the unified concept of energy—as something real yet immaterial—a weapon against materialism.

The Puritan Vision

It was implicit in Protestant theology that the mind of God is reasonable, commensurable with human rationality. Puritans, especially, favored reason, esteeming mathematics and physics as studies leading to the appreciation of a rational God.[34] They associated idle contemplation and blind faith with Catholicism, and associated the manual labor of experiment with the industriousness they valued. To them, science represented an opportunity to actively work for the benefit of mankind and the glory of God. Mechanism was accepted as founded on reason, precluding the heresies of pantheism, alchemy and astrology.[35] Puritans believed in predestination, a view entirely compatible with determinism and immutable laws of nature.[36]

The Enlightenment virtuosi were exclusively men.[37] They viewed matter as *materiel*, a raw resource for manufacturing.[38] A majority of the members of the Royal Society were Puritans, a group arising generally from the merchant class, which believed in salvation by good deeds and hard work. Science and technology promised to increase their personal status, as well as the general well-being of society.[39] Though lip service continued to be given to the status of nature as God's creation, nature was being assimilated to human energies, which were then discovering whole new continents of raw materials.

Science represented a way to participate in the divine plan. It was aligned with social and spiritual progress alike, which could be tangibly measured by technological advance. A religious umbrella initially unified the study of nature, the pilgrim's progress, and political action. The scientific revolution also coincided with revolutionary movements in society: the Reformation, the English Civil War, the American and French Revolutions. These signalled a shift away from medieval hierarchies and expanded the rights of a growing mercantile class, codified in documents such as the Magna Carta and the US Constitution, which were expressly independent of the arbitrary whims of monarchy. The scientific parallel was a standardized method, an objective framework and forum for knowledge. The stable natural world envisioned served to endorse the conditions necessary for the burgeoning industrial economy.[40]

From another perspective, however, this "rebirth" of Europe was something less than a great political leap forward. It derailed a trend toward constitutionalism already underway in the medieval period. Europe seemed ever ready to regress to earlier forms of patriarchy, in the absolutism of dictators and the divine right of kings.[41] Industrialization benefitted the masses—but only incidentally and unequally, as a by-product of the increased wealth and power of elites. Mechanization magnified physical power, and was potentially a leveler of social differences. In the long run, however, it magnified differences of class and social power. Despite early visionaries of science like Bacon, the machine did not evolve unequivocally as an instrument of benign social planning. Industrialization served less the intent to exploit technology for general human benefit, as Puritanism had proposed, than as a pretext to exploit cheap labor for the benefit of entrepreneurs. The consequences of an overarching economic mechanism outweighed the apparent beneficence of literal machines. The Christian veneer grew ever more superficial in the context of industrialization.

Secularism and Religious Resurgence

The resurgence of religious fundamentalism in the United States—and the responses of scientists to creationism—should be understood against a background of historical continuity and common ground between religion and science. Given the pivotal influence of Christianity on the development of science, resurging antagonism between religious and scientific communities should be assessed in the context of shared assumptions and values. The first scientists, after all, were unblushing creationists! Even in late nineteenth century, many scientists took an interest in spiritualism, with an idealistic regard for science as an ethical pursuit.[42] Some even hoped to scientifically demonstrate a spiritual basis for material reality.[43] Some contemporary Christians have come to regard the "new physics" as an ally of their faith, hoping to undermine what they perceive as the materialism of the Newtonian worldview.

Though the majority of modern scientists may not believe in the biblical God, or take interest in theological issues, they have inherited a tradition of thought that assumes the natural world to be a literal artifact, lacking intrinsic reality of its own. While science appears to survey the natural world from a materialist perspective, a major aspect of its approach is idealist if not outright theological.[44] It draws upon the Pythagoreans and Plato, as well as upon the heritage of Greek rationalism generally, which would reduce all knowledge to axiomatic systems. Today this thread is reflected in the perennial expectation that scientific knowledge is on the verge of completion, in a grand unified theory, and in the notion that the essence of physical reality is ultimately *non*physical, residing in a quasi-material vacuum or a completely nonmaterial substratum such as mathematics, computation, or information. Modern physical theory expresses an ancient skepticism in regard to the reality of the world. At the same time, it updates a perennial optimism that nature is not a mystery one simply comes upon, but is exhaustively knowable by virtue of its rational design.

Yet the gains of science in rational understanding were traded against a sense of being at home in the universe we try to understand. As one modern cosmologist famously observed: "The more the universe seems comprehensible, the more it seems pointless."[45] In other words, the world no longer seems divinely tailored to human meanings. One must ask, however, would a *scripted* world be better, with no place either for human freedom or an autonomous nature? To turn the question on its head, why should nature seem alien to us? While possibly religious, the question is also deeply psychological, political, and perhaps gendered as well.

The Enlightenment conceived the possibility of a humanly-created rational and secular order, a predictable "system of the world" that offered fulfillment through reason, technology, and enterprise. Gradually the conviction grew that industry and the state, rather than religion and morality, could guide society toward the equitable well-being promised by technologies of mass production. In many ways, however, modernity has failed to fulfill the social dreams of the founding fathers of science.[46] This may be one reason why religion continues to be surprisingly resurgent the world over, as the failures of secularism continue to unfold.

Religion reflects a deeply ingrained longing for moral perfection. Along with our suspect animal heritage, it denies the merely human capacity for social and moral perfection. For to take the beast out of its natural setting hardly removes the beastliness within. Failure to manage history and achieve a just society only mirrors failure to overcome those features of human nature that work against our happy coexistence with each other and with the natural world.

The resurgence of Islamic fundamentalism can be understood, in part, as a reaction to political errors of the West, as well as to the moral and social failures of capitalism and the global secular culture. Ironically, Christian fundamentalism is often so thoroughly integrated with capitalism that it has no coherently Christian social platform, as often demonstrated by the "religious right" in America. This failure owes something to the historical role of Christianity at the very heart of modernity. For a world that was *culturally but not ethically* Christian was essential to the arising of science, technology, and commerce.[47]

Christian doctrine endorsed the domination of nature and sanctioned the worldly expression of human will and masculine dominance, so long as it was nominally in divine service.

One motivation for religious faith is the quest for certainty—always resurgent in uncertain times. Religion provides, among other comforts, an axiomatic system of thought—a goal with which the theoretical physicist might sympathize. Its "theorems" are ideal in both senses of the word—representing a striving for perfection, while promoting ideals as already actualized truth. The closure of the system protects it from the interference of outsiders and of a reality that may be inherently unknowable and beyond control.

Religion has changed the face of the world, if not its personality. Sadly, on the large scale it manifests the same life cycle and processes of adaptation, corruption, and ossification that have defeated other great idealist movements, such as communism. Like the rest of society, it has been tainted by patriarchy, hierarchy, and lust for power. Secular culture embraced the hope that social measures could replace the inspiration and guidance provided by religion, yet its ideals, too, were systematically compromised. Education became purely intellectual, art devolved into decorative commodity, science became technological manipulation and philosophy became a technical exercise; literature was reduced to gratuitous storytelling, and politics is now little more than a tool of the rich to enslave the poor.

Notes

1. Davies, Paul. 1995. *Are We Alone?* New York: Basic Books, p. 96.
2. Dewart, Leslie. 1969. *The Foundations of Belief.* New York: Herder & Herder, p. 55.
3. Ibid, pp. 155 and 163. The Arabic translations of Aristotle, for example, introduced a distinction between two quite different meanings of *to be* not present in the Greek.
4. Nasr, Seyyed Hossein. 1996. *Religion and the Order of Nature.* Oxford: Oxford University Press, p. 174. The young Michelangelo was privy to humanist discussions in the household of the Medici.
5. Ashworth, William B., Jr. 1986. "Catholicism and Modern Science." In *God and Nature: Historical Essays on the Encounter between Christianity and Science*, edited by David. C. Lindberg and Ronald L. Numbers. Berkeley: University of California Press, p. 138.
6. Roger Hahn, Roger. 1986. "Laplace and the Mechanistic Universe." In *God and Nature*, edited by Lindberg and Numbers, pp. 265–66.
7. David Cayley, interview. 2010. "The Origins of the Modern Public." CBC *Ideas* series. www.cbc.ca/player/Radio/Ideas/Full+Episodes/2010/ ID/1528896270/. Protestants no longer viewed the Church literally as God's

administration on earth but as a fallible human institution. Such a shift in doctrine was essential to the reign of Henry VIII, for example, as titular head of the Church of England, rather than the Pope.
8. Torrance Kirby, interview. 2010. "The Origins of the Modern Public."
9. Barbour, Ian G. 1990. *Religion and Science: Historical and Contemporary Issues*. New York: Harper, p. 20.
10. Osler, Margaret. 1996. "From Immanent Natures to Nature as Artifice: The Reinterpretation of Final Causes in Seventeenth Century Natural Philosophy." *Monist* 79:3.
11. Lindberg, David C. 1992. *The Beginnings of Western Science: The European Scientific Tradition in Philosophical, Religious, and Institutional Context, 600 BC to AD 1450*. Chicago: University of Chicago Press, p. 150.
12. Grant, Edward. 1986. "Science and Theology in the Middle Ages." In *God and Nature*, edited by Lindberg and Numbers. pp. 58–59.
13. Jammer, Max. 1997. *Concepts of Mass in Classical and Modern Physics*. Mineola, NY: Dover, pp. 5 and 30.
14. Deason, Gary B. 1986. "Reformation Theology and the Mechanistic Conception of Nature." In *God and Nature*, edited by Lindberg and Numbers, p. 180.
15. Margaret Cavendish, Duchess of Newcastle-upon-Tyne (1623–1673), an early feminist writer and critic of the new science.
16. Menuge, Angus J. L. 2003. "Interpreting the Book of Nature." *Perspectives on Christian Faith* 55:90.
17. Barbour, *Religion and Science*, p. 26. Such pantheism was perceived as blasphemy and may have been one reason for the crackdown of the Inquisition.
18. Nadeau, Robert, and Menas Kafatos. 1999. *The Non-Local Universe: The New Physics and Matters of the Mind*. Oxford: Oxford University Press, p. 8.
19. Webster, Charles. 1982. *From Paracelsus to Newton: Magic and the Making of Modern Science*. Cambridge, UK: Cambridge University Press, p. 10.
20. Lindberg, *The Beginnings of Western Science*, p. 212.
21. Barbour, *Religion and Science*, p. 4.
22. Freeman, Charles. 2005. *The Closing of the Western Mind: The Rise of Faith and the Fall of Reason*. New York: Vintage.
23. Merton, Robert K. 1938. "Science, Technology, and Society in Seventeenth Century England." *Osiris* 4:429.
24. See my earlier book, Bruiger, Dan 2006. *Second Nature: the Man-made World of Idealism, Technology, and Power*. Trafford/Left Field Press.
25. Lindberg, *The Beginnings of Western Science*, p. 226.
26. Prigogine, Ilya, and Isabelle Stengers. 1984. *Order Out of Chaos: Man's New Dialogue with Nature*. New York: Bantam, p. 47.
27. Quoted in Lindberg, David C. 1986. "Science and the Early Church". In *God and Nature*, edited by Lindberg and Numbers, pp. 27–28.
28. Lindberg, *The Beginnings of Western Science*, pp. 237–39.
29. Merton, "Science, Technology, and Society in Seventeenth Century England," p. 417.
30. Merton, "Science, Technology, and Society in Seventeenth Century England," pp. 420–22.
31. Bono, James J. 1995. *The Word of God and the Languages of Man: Interpreting Nature in Early Modern Science, Vol. 1: Ficino to Descartes*. Madison: University of Wisconsin Press, p. 236.

32. Barbour, *Religion and Science*, p. 21.
33. von Baeyer, Hans Christian. 1999. *Warmth Disperses and Time Passes: The History of Heat.* New York: Modern Library, p. 114.
34. Merton, "Science, Technology, and Society in Seventeenth Century England," pp. 425–28.
35. Barbour, *Religion and Science*, p. 145.
36. Merton, "Science, Technology, and Society in Seventeenth Century England," p. 468.
37. While there were female natural philosophers, such as Margaret Cavendish, they were excluded from official bodies such as the Royal Society.
38. Merton, "Science, Technology, and Society in Seventeenth Century England," p. 506.
39. Merton, "Science, Technology, and Society in Seventeenth Century England," p. 440.
40. Jacob, Margaret C. "Christianity and the Newtonian Worldview." In *God and Nature*, edited by Lindberg and Numbers, pp. 240–44.
41. Franklin, James. 2001. *The Science of Conjecture: Evidence and Probability before Pascal.* Baltimore: John Hopkins University Press, p. 71. In this context, one must wonder at the role today of the quasi-divine stature of corporations in international law.
42. Powers, Jonathan. 1982. *Philosophy and the New Physics.* London: Methuen, p. 58.
43. Stewart, Balfour, and Peter Tait. 1873. *The Unseen Universe*, p. 5. Quoted in Powers, *ibid*, p. 89.
44. Paul Davies, 1995 Templeton Prize Lecture.
45. Weinberg, Steven. *The First Three Minutes*, New York: Basic Books, p. 154. In fairness, he adds, "But if there is no solace in the fruits of research, there is at least some consolation in the research itself. . . . The effort to understand the universe is one of the very few things that lifts human life above the level of farce and gives it some of the grace of tragedy."
46. See von Werlhof, Claudia, 2011. *The Failure of Modern Civilization and the Struggle for a "Deep" Alternative: On the "Critical Theory of Patriarchy" as a New Paradigm.* Frankfurt: Peter Lang.
47. See Dewart, *The Foundations of Belief,* p. 115.

13

Deductionism,[1] or the Proof Shall Make You Free

> *"The fit between reason in our minds and in the world would be expected if the world is the creation of mind."*
> —Ian Barbour[2]

> *"Never express yourself more clearly than you are able to think."*
> —Niels Bohr

Introduction

Though inspired by general expectations of how things behave in the real world, the fundaments of deductive systems are *true by definition*. One can know with certainty what a deductive system contains because it has been defined to contain those things. As with a computer program, one knows in advance that there is a specified output to any input. In contrast, in order to infer what a natural system contains, one must either take it apart or else observe its outputs in relation to inputs. Such inferences gain credibility by imagining the system to be a deductive system.

The laws of nature have a dual status: as synthetic generalizations and as analytic truths. On the one hand, they succinctly express results of observation and experiment. Such inductive generalizations cannot be proven, however, only disproven.[3] Theorems of mathematics, on the other hand, can be proven because they are propositions within formal systems, not empirical assertions. Whatever power of necessity natural laws appear to possess derives from translating inductive generalizations into theorems of a deductive system. Concerning physical phenomena, however, there are always alternatives to any proposed explanation.[4]

As we have seen, "natural necessity" boils down to logical necessity and "natural kinds" are a matter of definition. This is not to deny that there is something objective to which laws and kinds refer. But the role of the subject inevitably introduces ambiguity into knowledge. The important point here is that this ambiguity can be compensated by substituting precisely crafted definitions for the ambiguities found. Hence, the tacit scientific ideal is to translate inductive findings into deductive truths.[5] Laws of nature then become the rules of a game.

This ideal bears the promise that even apparently irreducible elements, such as basic physical constants, will one day be derived from a fundamental theory incorporating a handful of basic tenets. The "rules" would be specified by the theoretician rather than simply imposed by nature as irreducible empirical data. This project harks back to Greek deductionism and to Renaissance debates on the limits of divine power in setting the details of nature. The modern theoretician might prefer to derive all givens from first principles, leaving nothing to accident let alone to divine whim. However, scientific knowledge would then be empty formalism.

The development of logic and axiomatic systems by the Greeks reflected the social freedom to posit and accept given assumptions and rules of thought. At the root of deductive knowledge, therefore, lay the fundamental intentionality behind communication, reason, and argument: the free creative act of decree and whatever purposes inform it. Medieval thinkers inherited this rationalism, to which they added the belief that the universe is rational because a rational God created it. In the late medieval period, the task of theology was to systematize and clarify faith as a coherent body of propositions thereby rendered certain and final. Both Greek and religious influences thus favored deduction over induction, first principles over observation.

The syntactical aspects of language constitute a kind of familiar deductive system. Typically, nouns stand for real things, verbs for real actions, and grammatical operations stand for real features, relationships, or processes. Yet, in their own right, elements of language—as opposed to sematic content—are purely formal, without necessary reference to anything in the real world. Language is creative precisely because it can posit arbitrary forms, giving expression to imagination and even nonsense—as in the celebrated "Jabberwocky" verse by Lewis Carroll. Rhetoric can be persuasive because logic seems to hold together even when the conclusion seems absurd. The dependency of thought upon language may have particular consequences in the sciences, where the goal is to clarify what is real and semantically referential. As syntax can upstage semantics, so mathematical formalism can dominate the interpretation of nature—a possibility against which Bacon had cautioned.

Before 1500, vernacular language was not an object of study, much less something to formalize. One was simply immersed in it like water. Self-consciousness about language may have been a prerequisite for inquiry about nature.[6] The ability to study language objectively, and to reconstruct it in artificial versions, led to a similar approach to the Book of Nature. A natural language must be studied to discover its governing rules, to determine whether it *has* a grammar that can then be formalized. This modeling of language served as a paradigm for the modeling of nature, which was, so to speak, expressed in God's native tongue. The project of reconstructing language in a formalized version inspired the reconstruction of nature in formal scientific models.

Christian theology assumed that the mind of God is both manifest in nature and commensurable with human rationality. Without these assumptions there would have been little incentive to study nature. Yet skepticism in regard to the limits of human knowledge was still possible. For though God created the natural world, perhaps he alone could comprehend it. Perhaps divine whim outweighed divine rationality. Perhaps human rationality could not apply outside the human realm. Moreover, if people have available to them only concepts that spring from their own minds with which to study nature, perhaps these concepts—and the language in which they are expressed—are more appropriate objects of study than nature itself. Such reasoning early led Vico and others to the notion of "maker's knowledge."

While scientific thought does use ordinary language, it also creates its own novel symbols and jargon, giving rise to new forms of expression. Scientific terms differ from ordinary language insofar as they are precisely defined, free from intuitive secondary meanings or associations. In contrast, ordinary words are ambiguous, if only because they are tied to references in the complexly interrelated social world. The great advance represented by formal thinking was to define things precisely and unambiguously from scratch.[7] Formal scientific concepts were to mean only what they were explicitly held to mean. Words in ordinary language represent found things; terms in scientific language represent constructed meanings. The advantage of such definition is that one always knows what one is talking about; the disadvantage is that one can talk of nothing else.

Deductionism is the belief that physical processes correspond to such constructed meanings. In the extreme, it is the premise that nature itself is reducible to a mathematical model. Moreover, such models *are the actual objects of study*. Deductionism, in this sense, is an example of what Gerald Holton called *thematic content*: an implicit intuitive assumption. So is the theme of general mathematization that is the watershed of the Scientific Revolution. And so is the contemporary metaphysical belief that physical reality essentially *is* mathematics.

We have seen that the isolated system, determinism, reversibility, and equilibrium are byproducts of a mechanistic view of nature. They are properties of deductive systems. Mechanism reduces nature to artifacts (models, equations). Deductionism holds that physical systems *are* such artifacts—hence deterministic, reversible, etc. Properties of *deductive* systems are thereby ascribed to *natural* systems. While useful in many ways, this belief may exclude consideration of other properties than those defined in the deductive system (for example, properties of self-organization). If nature is real, not all its properties can be formally represented within such a system.

Vico: "The True is the Made"

Giambattista Vico (1668–1744) articulated the notion of *maker's knowledge*. This is the principle that we can understand best the things we make—whether conceptual models, literal machines, human institutions, or other artifacts. It represents the insider's knowledge the creator has of the creation. Applied to nature, obviously only God can have such knowledge. In contrast there should be no such limit on what people can know of *human* creations, such as history and philology. Hence, knowledge of man-made systems represents the most reliable knowledge. (From our present point of view, this is because it is effectively true by definition.) Vico argues that because humans know best the things they make, the most appropriate objects for study are cultural institutions and practices, as opposed to the natural world. Along with geometry, language, and other systems of signs, this would include history, politics, morals, and anything that is clearly a human creation.[8] Thus, while Bacon became known as the father of science, Vico is acknowledged as the founder of the social sciences.

One side of Vico's program is that things made by God remain incomprehensible, that one simply cannot fathom the divine mind as the early scientists hoped. Their argument had been that nature could be understood through a kind of reverse engineering, based on the belief that natural things are made by a rational God. Vico turns the argument around: we should not presume to know the mind of God through the study of nature, because we lack the Creator's direct knowledge of the created. On the other hand, we can at least hope to understand human institutions.

Maker's knowledge contrasts with contingent knowledge gained through interaction, observation, and use; it is knowledge of the made as opposed to the found. It includes the awareness of one's own intentions, as the designer of something, in contrast to knowledge of natural things inferred through second-guessing their functioning and parts. In a sense, Vico's principle turns the usual premise of first-order science upside down. Knowledge of nature, even through the scientific method, cannot be held to be reliable on the grounds that it is "objective." Rather, it is knowledge of our own "subjective" thought processes and concepts that should be considered reliable because these are what we can intimately know. Hence the significant place of ethics in Vico's thought.[9]

The principle of maker's knowledge expresses the allure of deductive systems. The idea that we have privileged knowledge of intentionally produced things could be called the "weak" maker's knowledge principle. In Vico's time, however, *everything* was understood to have been intentionally produced—when not by man, then by God. Vico's distinction had to do with the human versus the divine creator. (The idea of nature as self-existing, self-created, or having an autonomous reality was simply heresy in the Christian imagination.) A stronger version of Vico's dictum—*verum ipsum factum* (the true is the made)—draws upon the distinction between natural and artificial: we can know with certainty *only* our own constructs, in contrast to the found world. Deductive systems (such as arithmetic, geometry, and algebra) are the only systems, therefore, of which one can have guaranteed knowledge. A priori truths are not deep intuitions but tautologies—true by definition. Scientific knowledge is certain only to the extent it is knowledge of deductive systems. While nature may not be perfectly knowable, science as a humanly defined construction ought to be.

In our era, a possible implication is that the *design stance*, while a useful tool of science and human thought generally, may obstruct understanding the true complexity of natural systems.[10] For if nature is *self*-designing, its processes might be very different from those imagined by human designers.

Redefining Nature

Reflecting the divine power of the Logos, human creativity resides in the power to invoke clear and precise existence by decree or fiat. Nature, in contrast, is just what it is, independent of human declarations. It dwells outside our definitions and remains fundamentally ambiguous. Since we did not make it, we cannot know it with absolute certainty or finality. We *can*, potentially, have perfect knowledge about what we make. A fundamental human power over nature, then, is the power to redefine it in a conceptual system.

Scientific modeling reframes natural reality as consisting of theoretical entities and processes that are unambiguously defined. Poorly perceived or understood things are named, formalized, and treated as though definitely known. A model substitutes well-defined elements for the ill-defined found world. Yet, apart from the degree of correspondence ("accuracy"), the model is an entirely different sort of thing from what it models.

Greek thought transformed geometry from a practical, empirical, and informal system of measuring the lay of the land to a formalized, self-contained system of theoretical propositions and rules for reasoning about them. It redefined the world in terms of idealized objects, shapes, and logical relationships, reconfiguring space as an abstraction. Euclid's geometry remains the paradigm deductive system.

Like geometry, mathematics at large generalizes, abstracts, and formalizes basic relations observed in the world, which are then *defined* as elements of a deductive system. Hence, it is by redefining the world literally in its own terms that mathematics so powerfully describes nature. This is appropriate for those properties and relationships so general and basic that we accept them as inherent in the structure of *any possible* world—hence, as "logical" rather than merely "physical" truths. Because ultimately they reflect empirical generalizations, however, the mere fact of having declared them true by fiat offers no guarantee that they apply in every instance to the physical world. Even in familiar realms, idealization is often misleading.

Deductionism blurs the very distinction between mathematics and the physical world. While there is a natural continuity between logico-mathematical principles and physical laws, mathematics is at the extreme of generality concerning the behavior of things, whereas physical laws are more conditional and restrictive. However, one absolute difference is that physics is always contingent on (new) empirical evidence, while mathematics is always a product of definition.[11] It is indeed amazing that nature often does correspond to mathematics outside the original context. This does not mean, as some have surmised, that the world is itself a "mathematical structure," in some sense Platonic, but rather that we habitually seek out aspects of reality that can be *defined* as elements of some deductive system. It should not be so surprising then that these are commensurable with other aspects of nature also conceived as elements of a deductive system. That is, the correspondence between mathematics and physics is actually a correspondence of one deductive system to another.

Formalization

Like a board game, a deductive system (or *formal axiomatic system*) has playing pieces and rules defining possible moves, as well as a space in which the play occurs. These are conceptual things, which are exactly and only what they have been defined to be. This gives such systems a self-contained, tautological character unlike the real world.

Empirical facts do not have the formal validity that makes the truths of logic or the theorems of mathematics seem irresistibly necessary, which is why science is essentially different from mathematics. Mathematics has the precision of definition, while science has only the precision of measurement. On the other hand, logical truths are devoid of new information for this very reason. The deduction is already latent in the premises. The solution to a computational problem is a more explicit form of information already given implicitly in the problem.[12]

Like the rules of a game, formal elements are well defined and true by definition. For example, perfect right angles and dimensionless points do not exist in nature, and their idealized versions are not simply refinements of their physical counterparts. Rather, they are original products of definition, exact in principle, and so can be manipulated in thought with total precision. This is the advantage exploited in digital computing, free of the cumulative errors of analog processes.

In formalization, intuitive and vague notions are replaced with precisely defined elements and operations. A procedure is clearly defined for arriving at conclusions on the basis of specified assumptions and agreed-upon rules of reasoning. That is, a method of proof is established. While this allows logical necessity to be distinguished from probability or mere belief, the downside is that conclusions can only be as valid as the premises behind them. A deduction that starts out on the wrong foot remains forever on it.[13]

A deductive system is self-contained and coheres purely by logical necessity. There is no reference outside itself until it is "interpreted" as a mapping of some portion of the real world—just as plane geometry, for example, can be applied to the physical properties of the earth's surface. Then its premises may appear to coincide with truths about the world, its logical structure to mirror the organization of nature. But such a formalism can also be viewed as purely self-contained, a game played according to arbitrary rules unconnected to reality. The game of Monopoly, for example, may be viewed both ways. It is a self-contained system, yet was crudely modeled on economic realities.

Hilbert's program in the late nineteenth century called for the formalization of mathematics and physics. These are significantly different tasks, however. Mathematics is already a deductive system; the task of proving its completeness and consistency met with frustration, as demonstrated by Gödel. On the other hand, to formalize mechanics, for example, meant first regarding its concepts as fictions, products of definition. While this had its uses, the intention fundamentally contradicts the empirical nature of science. For nature is not an axiomatic system.

Weak and Strong Deductionism

The idea behind reduction is that the whole can be understood in terms of its parts, or in terms of a more fundamental level. Natural parts, however, are ambiguous and somewhat a matter of discretion. Furthermore, if nature happens to be indefinitely complex, there could be an endless regression of parts within parts, with no "fundamental" level. On the other hand, if nature has irreducibly fundamental elements, how do we explain the existence and properties of this bedrock when explanation is understood in terms of further reduction? The properties of truly fundamental particles and forces can only be accepted on faith, axiomatically, as though reality were ultimately a matter of our definitions.

Deductive systems are effective to the extent that simplification and idealization are. Mathematical equations define toy systems, which serve to model selected aspects of the real world. The equations may be isomorphic with these model systems, since they express the same abstraction. But neither is strictly isomorphic to the real processes or systems they represent. The belief that nature is rationally comprehensible, however, amounts to assuming that it can be assimilated to such models. Though gleaned from nature in the first place, the rationality is ours, to which nature has no obligation to conform.

Deductionism tends to mask the complexity of the real world when it *presumes* simplicity or prefers tidy systems to potentially messy facts. It is easy enough to mistake the model for the reality, when one thinks of nature as literally consisting of idealizations of various sorts. The limits of the deductive approach are the other side of its merits: conclusions are only as good as the assumptions they are based upon. It also tends toward speculation beyond experimental verification—the very kind of scholasticism rejected by the early scientists. The mild version of deductionism acknowledges that scientific models and laws of nature are matters of convention. One has creative freedom to propose different models, yet the question of which model fits best must be resolved through observation and experiment. The striking fit of mathematics with nature is most plausible on the stronger version of deductionism: that physical systems simply *are* deductive systems. Yet this is no more than a metaphysical fancy, which gains credence in modern times from the successes of digital computation.

Deriving Fundamental Constants

Some physicists believe the fundamental physical constants should be—and eventually will be—derived from theory.[14] That would imply, however, that the world is effectively a deductive system, and that the constants involved are merely constants of integration of the defining equations, or else conversion factors between different accounting systems. If nature is *real*, in the way I have suggested, then it will not be possible to derive *all* fundamental constants from theory; there will always be new ones to discover.

The universality of natural laws and the constancy of natural constants are matters of empirical observation, which means they are subject to revision in the light of new evidence. They are not eternal or absolute principles. Given the short history of science and the limits of observation, laws and constants of nature may be contingent and yet not change perceptibly over time or distance.

While some physicists might like every digit of every parameter to fall out from theory, physics would then be pure mathematics and not an empirical science. It would fulfill the Greek ideal, modeled on geometry, but not the Renaissance ideal based on observation and experiment. Yet the commitment to deductionism is strong enough that some theorists deny the reality of time, for example, in order to uphold the traditional idea of nature as a reversible (i.e., deductive) system whose fundamental description should be time-independent.[15]

Similarly, various theories attempt to reduce space to more "fundamental" abstractions such as information, geometry, or topological relations. Many scientists seem to believe that physical law exists in some sense independent of space, time, matter and energy—in other words, as axioms of a deductive system.

The Deductive Program

Euclid formalized geometry, which had been an informal and pragmatic set of observations about the world. He transformed what had been essentially an inductive enterprise into the deductive paradigm that has become the scientific ideal. In so doing, he made a qualitative leap from empirical generalizations to analytic truths.

Medieval thought had been essentially deductive, though mathematics then was not used to describe nature quantitatively so much as for theological argument.[16] Mathematics in the early modern period combined that tradition with equal measures of a quantitative approach to nature and Pythagorean number mysticism. The heritage of the ancients had impressed upon Christian thinkers the antique ideal of celestial perfection, so that Kepler tried to interpret the planetary orbits in terms of the five regular solids, and Galileo insisted on exactly circular orbits. Nowadays, such reasoning would be called argument from symmetry. While we may smile at the naiveté of earlier ideas, we have hardly outgrown the need to assimilate physical reality to idealized mathematical schemes.

Exact symmetry is a characteristic of idealizations, of products of definition. Things in nature are only ever *approximately* symmetrical. Exact symmetry is a property of equations or geometrical figures, not of nature. It *appears* to be an attribute of nature when we consider systems artificially defined and isolated, free from any interactions with the rest of the universe that might "break" the symmetry of the equations. Symmetry breaking as an explanatory principle, central in modern physics, attempts to account for this discrepancy between theory and actuality from a particular idealist bias. Yet by talking about "symmetry principles" as though they were a built-in fact of nature, one avoids mention of this bias and its human motivations. At heart, symmetry and symmetry-breaking are Platonic notions, holding that the phenomena we actually observe are but imperfect reflections of a deeper reality expressed by equations.[17]

This is hardly to deny the usefulness of symmetry arguments or of deductive reasoning generally in science. Indeed, Galileo came upon his famous discovery—that all bodies fall with the same acceleration—not by experiment, as commonly believed, but by simple reasoning in a thought experiment.[18] Thought experiments have played a similar creative role for many others—famously Einstein, who was one of the first to introduce modern symmetry arguments. Reliance on thought experiments, however, begs the question of to what extent nature can be presumed logical, and to what degree a general deductionist faith (such as Einstein's) is justified. The point here is not to limit the tool kit of science (as positivists proposed) but to caution against promoting tools—such as symmetry arguments—to the status of natural fundaments.

Descartes had set about to demonstrate that physical reality could be deduced from first principles.[19] Euler (1736) attempted to resolve logical problems involved with causality and force by axiomatizing physics.[20] Hilbert hoped to axiomatize the whole of mathematics. Gödel showed this was not possible by demonstrating *one* mathematical truth that evaded it. Chaitin showed that there are an infinite number of such truths. If mathematics as a whole cannot be formalized, how can physics be?

Determinism and Completeness

In his great debate with Bohr, Einstein's position is often considered to uphold realism (though actually it favors determinism) while Bohr's position is sometimes disparaged as subjective, positivist, or even idealist. However, the truth of the matter is rather the opposite. For determinism implies a deductive system rather than reality, while indeterminacy implies incommensurability with any deductive system—which is just what makes nature real rather than ideal.

For Einstein, the objective existence of physical reality implied deterministic parameters, perfectly knowable in principle. However, the reality of natural systems and the possibility of perfect knowledge are separate questions. Einstein's brand of "realism" is actually deductionism: the faith that nature can be understood unambiguously because it consists of well-defined elements. This is demonstrated by his insistence on the ideal of "completeness" in the famous Einstein-Podolsky-Rosen, or EPR, paper: *"Every element of the physical reality must have a counterpart in the physical theory."*[21] Such a formal mapping presumes that "every element" can be identified, which is possible only in a deductive system. If nature is real, however, no theory *can* be complete in this sense.

The probabilistic nature of nature, far from being an unsatisfactory state of affairs, is the very sign of nature's reality. Determinism, on the other hand, reflects the classical (and implicitly idealist) hope that nature can be reduced to a deductive system. Determinism, however, is not a property of the real world but of conceptual systems. Confusing them gives the illusion that the reality of nature can be completely represented. The significance of Bell-type experiments that have actually been conducted (physical versions of the EPR thought experiment) lies less in confirming quantum theory—which little needs further empirical confirmation—than in affirming that nature is *not* a deductive system. While a *deeper* level of analysis might be possible (as in hidden-variable theories), an *exhaustive* account is not. While Einstein may have been mistaken about quantum theory, and his quest for a unified theory was premature, his deductive approach has been widely embraced in modern attempts at a grand synthesis encompassing all the physical forces and entities.

Einstein's oft-quoted aphorism, "God does not play dice," expresses his distaste for the non-determinism and apparent non-realism of the quantum world. The subtext, however, is that he believed the world to be reducible to mathematical formalism.[22] Like Eddington, he hoped to reduce constants of nature to dimensionless numbers derivable from theory, on the belief that "in a reasonable theory, there are no (dimensionless) numbers whose values are only empirically determinable."[23] He felt that since the values of the fundamental constants ought to be derivable from theory, they would necessarily be computable numbers.

While not as literal as Newton, Einstein's reference to God should perhaps be taken as more than poetic. No doubt it owes less to any theological belief than to his faith in the rationality of nature. To a theist or Deist such as Newton, randomness expressed divine whim, which could not be second-guessed. But to a scientist without need for the God hypothesis, randomness in nature is simply a brute property of the world at a certain level. Einstein's distaste for such a property reflects his implicit rationalism, based on the hope that no aspect of nature should elude analysis. In his wake, modern practices—such as symmetry arguments and adjusting parameters of computer models—are based on the faith that nature itself is organized as an equation, program, or deductive system.

Dreams of Reason

Following Hume (and Piaget), one concludes that the only plausible meaning of causal determinism is logical implication; hence, *the only truly deterministic systems are deductive systems.* Following Vico, one concludes that *the only certain knowledge concerns models*. While equations allow us to predict the behavior of models, there is no guarantee that the model effectively represents the reality. If it *happens to*, within limits, this is an empirical fact, which cannot be assumed a priori. Contrary to the dream of reason, knowledge of nature is always contingent and essentially probabilistic, even on the macro scale. Physical knowledge pertains to *phenomena*, in Bohr's sense, rather than to unambiguous realities, in Einstein's sense. Though differential equations may perfectly describe an idealized system, in the end they only correspond to reality as well as they correspond statistically to empirical data. The equations appear to *define* the reality, but in truth they define the model and only approximate the data. While nature cannot lie to us, the laws of physics can.[24]

Some early commentators, among them Herschel, had recognized the limits of the deductive program. Even the great theorist Maxwell expressed skepticism, acknowledging that dynamics deals with bodies that have only the properties formally assigned to them.[25] Einstein, on the other hand, held to a more Platonic vision, with confidence that "pure thought can grasp reality, as the ancients dreamed."[26] Eddington expressed similar views. Though the efforts of these luminaries to derive everything from theory—including the fundamental constants—were fruitless in their day, others have renewed them in recent times. Yet the very division of theory into laws and initial conditions reflects the fact that there is always a remainder, outside theory, of details that are exceptions to the rule. If nature is *real*, then the theorist's dream of a complete and final theory is illusion.

The deductionist tradition going back to Greek idealism had been transmitted through Galileo, for whom the language of nature was mathematics. It was reinforced by Descartes, for whom reality was but extension in space. It permeated the rationalism of Christian Europe as what some have called "the hidden ontology of classical epistemology," which is alive and well today.[27] It flowered in great religious minds, like Newton's, and in great secular minds like Einstein's. It continues to be reflected in the neo-scholasticism of the modern string theorists and cosmologists. It shows up in society's expectations for quantum computers, and in the hopes of some mathematicians concerning computation in polynomial time.[28] It is expressed in the views of some contemporary mathematicians and physicists alike, that the universe is fundamentally made of numbers.

Notes

1. I choose the term "deductionism" over "deductivism" to differentiate the intent here from other uses of the latter term. While deductivism refers to a process of modeling that reduces appearances to the elements of a deductive system, I mean deductionism to refer also to the implicit belief that nature actually corresponds to such deductive systems—not only "for all practical purposes," but qualitatively. In its extreme, deductionism is the belief that models work because they are identical to natural realities they model.
2. Barbour, Ian G. 1990. *Religion and Science: Historical and Contemporary Issues.* New York: Harper, p. 211, discussing ideas of John Polkinghorne.
3. There is not only a logical distinction, following Hume and Popper, but also a psychological distinction between the certainty of logical proof and the reasonable probability of evidence upon which much of actual science depends, as in everyday life.
4. Wick, David. 1995. *The Infamous Boundary: Seven Decades of Heresy in Quantum Physic.* New York: Springer-Verlag, p. 173.
5. Cf. Earman, John. 1986. *A Primer on Determinism* Dordrecht, NL: D. Reidel, p. 88: "In much of the current literature on the structure and function of scientific theories, 'theory' and 'deductive system' can be freely interchanged."
6. Roland Greene, interview. 2010. "The Origins of the Modern Public." CBC *Ideas* series. www.cbc.ca/player/Radio/Ideas/Full+Episodes/2010/ID/1528896270/.
7. Cassirer, Ernst. 1980. *The Philosophy of Symbolic Forms, Vol. 3: The Phenomenology of Knowledge.* New Haven, CT: Yale University Press, pp. 334–40.
8. Danilo Marcondes de Souza Filho "Skepticism and the Philosophy of Language in Early Modern Thought" Pontifícia Universidade Católica do Rio de Janeiro, Brazil.
9. Koerner, Stephanie. "The Status of Ethics in Contemporary Epistemology and Ontology, and the Problem of Meanings and Values (the Symbolic) in Archaeology." The Criteria of Symnolicity. www.semioticon.com/virtuals/symbolicity/status.html.
10. The design stance is the idea popularized by philosopher Daniel Dennett that it is useful to regard natural systems as though they had been designed.
11. Of course, there can always be new definitions. On the other hand, new physics has sometimes required new mathematics.
12. Goldreich, Oded. 2005. "Randomness and Computation." www.wisdom.weizmann.ac.il/~oded/r+c.html.
13. While there is no accumulation of error in digital processes, everyone has experienced the kind of catastrophic corruption a damaged DVD can display.
14. The basic "fundamental" constants are charge, mass, and magnetic moment of elementary particles. Next in "primitiveness" are c, h, G, and k. All other important universal constants are based on these seven. See Johnson, Peter.1997. *The Constants of Nature: A Realist Account.* Farnham, UK: Ashgate, p. 4.
15. See, for example, Barbour, Julian. 1999. *The End of Time.* Oxford: Oxford University Press.
16. Lindberg, David C. 1992. *The Beginnings of Western Science: The European Scientific Tradition in Philosophical, Religious, and Institutional Context, 600 BC to AD 1450.* Chicago: University of Chicago Press, p. 203.

17. Weinberg, Steven. 1992. *Dreams of a Final Theory.* New York: Pantheon, p. 195.
18. Hamming, R. W. 1980. "The Unreasonable Effectiveness of Mathematics," *American Mathematical Monthly* 87:81–90. Specifically, he reasoned about dividing a falling body into smaller (lighter) parts, which should then fall at different rates if the acceleration depended on mass.
19. van Fraassen, Bas C. 1999. "Structure: Its Shadow and Substance." PhilSci Archive. http://philsci-archive.pitt.edu/631/1/StructureBvF.pdf.
20. Jammer, Max. 1957. *Concepts of Force.* Cambridge, MA: Harvard University Press, pp. 211–12.
21. Einstein, A., B. Podolsky, and N. Rosen. 1935. "Can Quantum-Mechanical Description of Physical Reality Be Considered Complete?" *Physical Review* 47:777–80.
22. Barrow, John D. 1991. *Theories of Everything: The Quest for Ultimate Explanation.* New York: Fawcett/Balantine, p. 244. According to Barrow, Einstein's success with mathematics in developing the general theory of relativity led him to abandon his earlier simplicity of thought in favor of mathematical formalism.
23. Quoted in Rosenthal-Schneider, Ilse. [Einstein's words as quoted by Rosenthal-Schneider in her chapter of the book] In *Albert Einstein: Philosopher-Scientist.* Vol. 1, edited by P. A. Schilpp, New York: Harper Torchbooks, p. 144.
24. Cf. Cartwright, Nancy. 1983. *How the Laws of Nature Lie.* Oxford: Oxford University Press.
25. Morrison, Margaret. 2000. *Unifying Scientific Theories: Physical Concepts and Mathematical Structures.* Cambridge, UK: Cambridge University Press, p. 101.
26. Einstein, quoted in Holton, Gerald. 1996. *Thematic Origins of Scientific Thought,* Cambridge, MA: Harvard University Press, p. 252. In childhood, Einstein's first mentor gave him the "little holy geometry book" (an exposition of Euclid). Like the magnetic compass, this made a big impression on him and a lasting influence on his style of thought. Geometry was to become an enduring theme in the content of his work, as well as a model for his approach.
27. Nadeau, Robert, and Menas Kafatos. 1999. *The Non-Local Universe: The New Physics and Matters of the Mind.* Oxford: Oxford University Press, p. 8.
28. Cf. Gasarch, William I. 2002. *SIGACT News* 33:34–47. It seems intuitively clear that P=NP could be true only in a simulation or in discrete mathematics. The question of P vs. NP (explored in the film *Traveling Salesman,* for example) therefore boils down to the question of whether the natural world is real.

14

Ideality

"Smooth shapes are very rare in the wild but extremely important in the ivory tower and the factory."
—Benoît Mandelbrot[1]

"Simplicity, like everything else, must be explained."
—Steven Weinberg

Scientific Aims and Assumptions

Science assimilates the complexity of appearances to relatively simple models. It does this by restricting its scope to simplified situations, with a minimum of factors in play. The virtue of such abstractions is that they can be treated mathematically, which means that parameters of particular interest can be isolated and predicted, though at the cost of omitting others.

Models are proposed according to specific explanatory aims, which may be more adequate for some purposes than others.[2] Scale partially defines what counts as data and as structure. Elements are isolated from their natural context and specific behaviors of interest are identified. What is excluded from such definitions and aims becomes a background "noise," which continues nevertheless to circumscribe the accuracy and scope of the theory.[3] And what is included rests upon assumptions that are not always explicit.

The founding fathers of science assumed, for instance, that nature is comprehensible and that matter lacks inherent powers of self-organization. Matter could be moved, but was incapable of moving itself. It consisted of fundamental, irreducible entities upon which "forces" acted externally. This way of thinking reflected the exclusion of all but efficient causes in the mechanist purview. Efficient cause carried with it an assumed dualism in which the chain of efficient causes must be initiated ultimately by a nonmaterial agent. Thus, the passivity of matter was built into how it is partitioned. The notion of a single particle in isolation, for example, gives the impression that it passively follows simple laws of motion. A different view would be that a particle is influenced in a very complicated way by every other particle in the universe.

Another tacit assumption can be seen in the treatment of space as geometrical, as illustrated in the familiar Cartesian coordinates. However, to render space (or time) geometrical is to redefine it as a deductive system. The laws of nature are then analogous to rules of logic, initial conditions analogous to axioms.[4] However, some axiomatic systems are complex enough to express propositions that cannot be proven within them. If *that* analogy holds, then at least some physical systems must be complex enough that their behavior cannot be derived from initial conditions through the known laws that pertain to them.

Science as we know it is possible because nature *can* be assimilated to simplistic models—at least "for all practical purposes." If this were not the case, there would be no advanced technology, which depends upon the effectiveness of such models. Yet while the philosophy of mechanism facilitates engineering, one may wonder how much it facilitates *understanding* of nature, in contrast to the goals of prediction, control, and use. This may hardly be an issue for many scientists, yet in the opinion of others there remains a grander enterprise of understanding the world.[5] Moreover, theory and practice have a reciprocal relationship, so that a deeper understanding may expand the sphere of practical purposes in ways not presently imaginable.

The guiding ideal of simplicity, still widely embraced as Occam's razor, is a metaphysical assumption with little empirical basis. Aside from human preference, there is no a priori reason why the world should be simple or why simple explanations should be true more often than complicated ones. It may be no more than a prejudice deriving from human cognitive needs and conditioning.[6]

All assumptions are at once enabling and limiting. Perhaps the most fundamental premise of physical science as we know it has been its outward focus on the natural world, which tacitly excludes the scientist. This is the presumption that the object can be separated from the subject, which is the defining premise of first-order science. The theorist's or experimentalist's input is off limits except as regards *physical* effects, such as the influence of apparatus on experimental findings or the relativistic effects of the observer's state of motion. The theorist may thus leave to philosophers of science the task of challenging shared assumptions. Yet the more basic and widely shared these assumptions are, the more likely they are to escape critical notice.

Besides natural simplicity and the outward orientation, other basic themes worth questioning include: the principle of *sufficient reason*, whereby everything is assumed to have a knowable cause; the *identity of indiscernibles*, whereby things do not simultaneously occupy the same position; *computerization* and *mathematization*, with the assumption that nature can be exhaustively represented; the search for *single causes* and the principle of *ceteris paribus*; the *specious improbability* associated with fine-tuning arguments; *reification* in general; assumed aesthetic guidelines for theory, such as *generality, symmetry, invariance, beauty,* and *elegance*; the assumption that presently accepted *categories* reflect real structure and that nature can, in general, be carved along its real joints; and various assumptions concerning *entropy, reversibility,* and *order*.

Idealization

To idealize is to remake the world according to taste—which, of course, means *someone's* taste. Idealization follows naturally from the ability to generalize, abstract, and extrapolate. It is conditioned by language and guided by preferences about how things should be. It is the basis of mechanism and deductive systems.

Ideals have both a prescriptive and a descriptive sense. The first indicates a standard of perfection—how things *should* be. The descriptive sense concerns the world of *ideas*. These two are intimately related, of course, since any notion of perfection is an idea, and various ideas serve as standards against which to evaluate what actually is. Ideals are typically projected as objective realities, since otherwise they might seem to lack the requisite normative force. Hence, Plato's idealism is at once descriptive and normative. In a tidy reversal of common sense, the ideal world he advocates as *desirable* is held also to describe what *truly* exists: an eternal and objective realm of perfection of which the actual world of appearances is but an imperfect and ephemeral copy. Aristotle disagreed with this view, holding that the reality of things resides in their individuality, rather than in some human way of categorizing them. He was more interested than Plato in the particularity of things. We can restate his view thus: reality lies in the ephemeral details more than in generalities or definitions.

Scientific thought attempts to account for the details *in terms of* the generalities: to express the found in terms of the made, the real in terms of the ideal, the natural in terms of mathematical constructs. The classical analysis of physics was framed in terms of differentiable functions (smooth curves). Aspects of nature chosen for study were those amenable to such analysis.[7] Less amenable aspects have since been discovered—for example, deterministic chaos and complex self-similarity.

Analysis involves the chunking of detail, a level of coarse graining. While fidelity is relative to the scale of graining, the unit of this texture is a matter of definition. In the mechanist metaphor, such units amount to parts of a conceptual machine—definite and formally specifiable. The world may or may not have such ultimate structure; it may have continuous *and* discrete aspects at every scale. The question of nature's intrinsic "bottom" must be distinguished from any discreteness imposed by theory. Similarly, the question of nature's intrinsic symmetries must be distinguished from defined idealizations, for there are no exact symmetries in nature, only approximate or "broken" ones.[8]

Natural Idealism

Natural or naive realism reflects faith in the literal truth of cognition. Through normal acts of projection, the brain's simulation of the external world is experienced *as* the world. Natural idealism is the complementary intuition that the true essence of the world consists in ideas, thoughts, or perceptions rather than material things. While appearing to be its opposite, natural idealism extends natural realism through processes of abstraction, reification, and projection. Both are concerned with what fundamentally exists and both produce their favored ontologies.

We have seen that one job of subjective consciousness is to reappropriate projected psychic contents, which are thus "de-realized," so to speak. Like realism, idealism does the reverse. It makes the ideal seem real, assigning to mental contents a status independent of the self and superordinate both to man and nature. Even so, the ideal is based on experience with the physical—although re-created in humanly preferred terms (which is the whole point). Idealization simplifies and abstracts reality, transcribing it into defined terms. It translates extension into intension. No idealization, however, can reflect reality in its fullness, nor even necessarily in its essence.

Materialism and idealism foster divergent cosmologies and causal histories. They have radically different ways to understand the present. Idealism typically inverts the causal relationships and temporal order of materialism. The idealist goal of knowledge is not to grasp the reality of nature but the illusory nature of ephemeral experience, in contrast to the eternal truth of the ideal—whether the forms of Plato or the eternal transcendent mathematical laws of nature.

Mathematical Objects

In contrast to physical objects, the objects of mathematics are timeless and nonphysical idealizations. They have been raised by fiat to this status, becoming products of definition. While change renders nature real, mathematics formalizes what are perceived to be changeless properties. Geometrical and logical concepts no longer refer directly to physical things or relationships, but to precisely defined abstractions. The embodied subject perceives a changing landscape, but mathematical laws and objects, like the invariance principles of physics, represent the search for constancy beyond change and apart from perceiving subjects.

To the degree that logic reflects, generalizes, and abstracts actual experience, it is to be expected that logical truth should correspond to physical truth. Yet there is no a priori guarantee of correspondence. On the contrary, there may be a selection effect involved, such that reality appears to correspond to mathematics simply because we focus on those aspects of the world we can treat mathematically—those that can be readily idealized.[9] A second reason for the mysterious correspondence is that mathematics is not actually compared to physical reality but to ersatz models, which are themselves idealizations. One deductive system can be commensurate with another because they are both products of definition. Mathematics can apply effectively in "unfamiliar" situations that have thus been redefined in familiar terms. In that sense, the correspondence is actually between one part of mathematics and another.

Mathematical truth is dual, both a free invention and a reflection of nature. To an undetermined extent, one can rely upon nature to follow known mathematical principles; but there may be areas where she does not, pointing perhaps to future mathematics. Like all forms of cognition, mathematics is a product both of the world and of the mind. The scientist does not circumvent this truism by pretending objectivity, nor does the mathematician by recourse to Platonism. Yet this does not prevent some mathematicians from believing that mathematics has a reality independent both of nature and of human minds.

While it may be *convenient* to treat mathematics platonically in some contexts, in others it could prove equally useful to consider mathematics a genetically and culturally constructed map of nature—that is, one that reflects both the physical world *and* the human mind's adaptive capacity to abstract. To reify mathematics as a platonic realm is to ignore both its genetically inbuilt relation to physical reality (the real preestablished harmony) and its significance as a conscious human creation. Some have argued that—because it transcends physical reality and particular forms of embodied cognition—mathematics is the plausible basis for communicating scientifically with an alien civilization. However, this ignores the tortuous history of mathematics on *this* planet and the cultural determinants of its present hegemony in the global monoculture. Mathematics may prove to be a more plausible basis for communicating with aliens about *technology* than about physical reality. On the other hand, like us, they might simply view physical reality through their technology.

Galileo claimed that the Book of Nature is "written in the mathematical language, and the symbols are triangles, circles, and other geometrical figures. . . ."[10] Yet pure geometrical figures are certainly *not* literally evident in nature. It puts the cart before the horse to assert that nature *expresses* mathematics, when mathematics was devised to express idealized natural relationships. Galileo's assertion reflects the neo-Platonism of the day, for Plato had held the basic elements of physical reality to *be* geometric forms. Yet even the planets did not appear to move in the ideal circles Galileo supposed, nor uniformly, but in complicated advances and retreats. It was a great creative leap of imagination to assimilate these complex changes to simple laws and an idealized model of planetary motion, which we have come to accept as the reality of the solar system.

Nomological Machines and Monsters of Nature

Scientists understandably focus on things they can potentially understand—having created them themselves. One sees this tangibly reflected in the popularity at one time of "orreries"—mechanical models of the solar system—and in the current use of molecular models in chemistry. But scientific constructs also include diverse abstractions such as particles and fields, isolated or closed systems, mathematical models in general, geometricized space and time, controlled experiments, etc. Philosopher Nancy Cartwright generically calls such artifacts *nomological machines*.[11] Deductionism is the belief that all of physical reality can be mapped by such constructs, and is even isomorphic to literal machines such as quantum computers.[12]

A controlled experiment is such a "machine."[13] Like other machines, it has a design potentially realizable in physical materials in a real environment. Select factors are isolated for study, with other influences excluded. This is not the situation in nature, where the real system generally cannot be controlled to isolate specific factors of interest, and the number of factors involved is indefinite. In an experiment, ideally a single factor is studied. The idealization is justified on the ground that the gain of treating the situation mathematically outweighs any loss from simplification. The isolation of causes is justified on the ground that in nature some causal factors are overwhelmingly more significant than others—*for the desired purposes*. It is argued that science would simply not be possible if this were not so.[14]

Robert Boyle articulated the key role of experiment in the science emerging from natural philosophy: "artificial and designed experiments are usually more instructive than observations of nature's spontaneous acting."[15] This is directly contrary to Aristotle's admonition not to interfere in natural processes, which could only be properly understood by limiting human involvement to passive observation. Observation by itself, however, does not pose questions and is ultimately but a more subtle interaction (e.g., through an exchange of photons). Science often begins with a question, framed within current ideas; one must know what to observe that would constitute an answer. Experiment clarifies the situation: the answer is how the experimental setup responds to a defined input. However, this may or may not correspond to the question the scientist has in mind. The experiment may answer a different question, or indicate that the wrong question was asked in the first place. The advantage and the disadvantage are the same: the experiment redefines the question but also constrains the answer. "Free" nature might answer differently, which was Aristotle's point.

For Aristotle, what we now call experiment was anathema; it could only produce "monsters of nature"—unnatural situations that result from human interference and do not represent knowledge of what nature is in itself. A more neutral way of putting it would be to say that experiment is not a window on the world but an interaction of the experimental apparatus with it. We now recognize that all knowledge of the world must involve some such interaction, even when the "apparatus" is human sense organs. Yet we tend to think of experiment as directly revealing natural reality, in the way the senses do, and to give insufficient consideration to the role of the experimental setup in determining the answers to our questions. This is where Aristotle's caveat still holds value in our day.

One can specify the input to an equation or computer program; but one can impose initial conditions only approximately on real situations, such as a laboratory experiment or the launching of a rocket. The real situation is subject to physical forces not included in the equation or program that models the experiment or the trajectory of the rocket. The experiment may overlook unknown factors that should not be dismissed as insignificant. Specifying input is not possible in real situations in which it cannot be controlled in detail (for example, specifying the individual motions of molecules in a gas).

Laws of nature involve an act of faith, originally informed by religious and deductive traditions, and justified by success at prediction.[16] Laws generally describe ideal things and circumstances that do *not* occur in nature; they could not even be stated without idealization, isolation, and experimental "control."[17] To single out a causal relationship, one must know how the process would occur in the absence of other influences. In nature, however, there are always multiple influences, some of which are disregarded as irrelevant.

Such a strategy must presuppose its own conclusion: that the idealized or isolated system reflects "reality" because factors not included in the model are irrelevant.

The faith behind this strategy seems justified in many circumstances. Yet what if the indefinitely many factors involved in natural systems and processes were not considered irrelevant? Would that mean that no laws could be expressed? Would it mean that nature could not be treated mathematically, or only that more sophisticated mathematics is required? Perhaps the simplifications typical of classical physics, represented by linear equations, are no longer either sufficient or necessary for the progress of science. We are on the verge of glimpsing the underlying connectedness of everything to everything else; and the computer now enables us to begin to engage the complexities of this wholeness.

Models can depart from reality in two closely related ways. First, through how they are defined; second, through how they ignore factors that may affect the phenomenon in subtle or unacknowledged ways. Successful idealizations tend to become accepted "realities," covering the human tracks of how and why they were proposed. This makes it easier to forget that laws of nature are conventions, and that the role of experiment is to find *which* conventions best fit the data.[18]

Isolated Systems, Reversibility, and Equilibrium

Isolated systems, determinism, reversibility, and equilibrium are artifacts of a mechanistic view of nature. If the world happens to be a plenum, then every region of it must be interconnected, at every scale, with every event connected to all other events backward and forward in time, with varying immediacy. If all were of equal significance, prediction would be impossible—indeed, life would be impossible since we are shielded, if only by distance, from many influences that could be lethal. Prediction requires ranking significance, especially by drawing a line to define an isolated system. It requires determinism (which we have seen is effectively deductionism), for otherwise it would not be possible to predict the behavior of systems with only laws and initial conditions.[19] Mere physical isolation is not sufficient, for within even a limited volume there could be indefinite connectivity without determinism. The system must be *logically* closed—re-defined as a deductive system. (On the other hand, if the world happens to have an ultimate discrete structure, then it would have a finite number of possible interconnections, events, causes, and effects. The task then would be to understand this *physical* discreteness as a real phenomenon, apart from mathematically *defined* discreteness. If the world is ultimately discrete, it must be so for deeply physical reasons, not because it corresponds to current technology or guarantees computability.)

A system is physically closed when there is no material exchange with an environment. It is logically closed when no process within refers to anything outside its definitions, including a background against which to measure change. Then it appears time reversible because time is irrelevant in a deductive system. Equations are reversible, but not real systems. No real system (except, by definition, the universe as a whole) is logically closed or absolutely isolated; there is always an "outside" that constitutes a background that can impinge with multiple extraneous causes, and against which processes can appear irreversible and asymmetric.[20]

Closed reversible systems are the staple of classical physics because one can treat them mathematically. The fundamental laws of physics are generally "time-reversal invariant," even though many physical processes at the macroscopic scale seem irreversible, as expressed in the second law of thermodynamics. However, strictly speaking, this reversibility is not a property of nature but a mathematical property of equations ($-t$ can be substituted for t). One can calculate the behavior of the *model* backward or forward in time at will. Though time plays a key role within such a framework, it is irrelevant to the framework itself, since deductive systems are timeless by definition. In the real universe, however, there is a direction of time. There are irreversible processes because no part of the real world is actually a deductive system, or ruled by an equation, even though it may be convenient to treat it so. Irreversibility is fundamental even at the quantum level, as demonstrated by the "collapse of the wave function."

Dynamics is defined in terms of instantaneous time, and the passage of time required for a measurement is typically not taken into account. Quite apart from the uncertainty principle, however, the very fact of motion implies a "classical" uncertainty in measurements. Unless the world is fundamentally discrete, precise instants are a fiction, for there is change during any actual interval, however short.[21] Reversibility thus depends on a generally unrealizable definition of precise instants. Moreover, time is problematic in extreme conditions where nothing "happens" that constitutes observable change.[22]

A measurement is thought to reveal an objective property of a system if it can be repeated *under the same conditions* with the same result, "all other things being equal." As Heraclitus noted, however, in the real world all other things are *not* equal over time, since the world is in flux. The repetition of conditions can only be approximate, and the results of measurements are therefore statistical.

Environment is but one "background" against which a process may be judged irreversible. A system may also undergo "internal" changes unaccounted for in its defining equations.[23] Friction and deformation are classical examples, where objects cannot be treated as point masses in elastic collision. A reversible process, in contrast, assumes that all variables affecting the development of the system in time are expressed in the equations of motion that are taken to define the system. This works in enough cases that we are led to believe in the natural existence of reversible processes.

During the era of its stability, the orbit of a planet, for example, may be mathematically described as a function of time, by considering only its mass, position, and velocity and treating its motion as frictionless and reversible. However, the era of its formation involved irreversible processes, as will the eventual dissolution of the solar system. Orbital dynamics disregards the history of the solar system, internal changes of the planet, and background galactic and cosmic processes that may not significantly affect planetary motion over "long" periods of time.

Such things *do* affect the evolution of the system, however, when it is not simply *defined* in these exclusive terms for a particular epoch or relative to idealized space and time. Relative to a background of actual terrestrial, planetary, and cosmic history, planetary motion is obviously *not* reversible. It only appears so when measured against artificial, ahistorical reference frames.

The concept of a time-reversible system was probably inspired by special cases such as celestial motions and simple cyclical mechanisms, where reversing the direction of motion does not affect the overall behavior of the system as defined. Yet the notion of reversible physical systems leads to problems, such as the arrow of time or the problem of accounting for why events in the real world are in fact irreversible. For most natural processes—and even some devices—are not reversible in the above sense. A gun, for instance, cannot "un-fire." And to actually reverse the direction of motion of a planet would be catastrophic for it.

It should come as no surprise that one cannot move backward in time, since in truth one cannot retrace one's steps in space either. Motion through space seems reversible only when space is a static idealization with background change defined out of it. In such geometrically conceived space, "motion" (i.e., along some coordinate from an origin) is but an ordering of magnitude, an operation by definition reversible.[24] (One can count backwards or forwards at will.) Motion through physical space seems to coincide with this, in that a continuous progression seems possible from object to object or from perspective to perspective. But such an accounting is only reversible within an artificial reference frame (grid) that is assumed from the outset to be unchanging; you cannot retrace your steps in an evolving landscape, except by disregarding the aspects that are changing. In real space, landmarks move about, the perspectives and background change not merely as a result of one's own movement but also because they evolve on their own, so that it is never possible to re-occupy the *same* perspective. A metrical grid that changes in a similar way would render the idea of reversible motion within it very complicated, to say the least. Static isotropic space is an idealization, and some such idealization (a rigid reference frame) is necessary to study motion at all. But this does not mean that it corresponds to, let alone defines, real space.

While the notion of reversible mobility through space depends on an artificial reference frame, the *natural* reference frame is simply the world itself as perceived, if not in its entirety. It is not some arbitrary landmark or imposed grid. When location is defined in reference to an ever-changing world, as Heraclitus observed, a moving observer can never return to the origin, whose very meaning is constantly in flux. In fact, one cannot be at rest relative to a restless world, any more than one can stop change. The fundamental reality, then, is *change*. We have, by convention, artificially resolved "change" into "motion through space" and "the flow of time." *Time-reversal invariance* is thus ambiguous, since it is unclear what is reversed. It could be the time signature in an equation ($-t$ for t), or it could mean reversing the direction of real motions or the polarity of field strengths, for example. That could mean the difference between reversing a predefined trajectory in an abstract conceptual space and reversing the real motions of the planets or even of everything in the universe!

Making a movie of a dynamic "system" is often proposed as a test of its reversibility. If what you see looks "natural" when the film is run backwards, then the system is reversible. For example, if one filmed the motion of several billiard balls or of several gas molecules, the reversed motion seen in the film played backward would seem entirely plausible. Contrast that with filming a cup falling off a table and breaking into shards. However, if what is filmed includes a context (for example, off the billiard table or outside the container of the gas molecules), playing the film forward and backward is *not* symmetrical, since the background will have changed. Clearly playing a film of the breaking cup backwards does not look natural, for we don't normally see broken cups reassembling themselves. However, if we pay attention to the changing background of the billiard ball as part of its system, we don't see that system regaining an earlier configuration either. It is only by ignoring the background that the motion of the ball in the film seems similar played forward or backward. If we viewed only the motion of several molecules of the breaking cup in isolation, and ignored the state of the cup as a whole, their motion would similarly appear reversible.

Space and time are concepts imposed upon change, which is a fundamental aspect of phenomenal reality. Geometry was the first deductive system formalized, and clocks were among the first mechanisms—becoming the paradigm of the mechanical philosophy. There are, of course, relatively stable and fixed features of the solid terrestrial landscape, and there are highly regular celestial phenomena seen to mark change and able to measure it. Such natural references were abstracted to create artificial reference frames against which change could be measured. In this sense, the invention of the ruler and the clock coincided with the invention of space and time as formal concepts.

These became the foundation of physics, so that most equations are expressed as a function of a time variable, and/or position variables, and their derivatives.

The fact that we do not have mobility in time means simply that we cannot control or alter the ongoing natural change of the world. Nor do we actually have the mobility in space that we take for granted, in the sense that to truly return to the "same" place would be to undo the changes that occurred in the world during the course of the journey. One cannot traverse the "same" distance twice, if referents defining it are in flux. One can move reversibly only in a fictional static world, where apparent change is a function only of one's own movement. A system may appear reversible when measured against rulers and clocks that are thus *defined* to be stable. But wherever the system includes a natural environment, so that events are measured against a background of unpredictable change, the progress of events is irreversible.[25] Intuitively, one recognizes that the course of real events, though measured by the clock, refers ultimately to background events in the rest of world, for which the clock stands in. Clocks may be made to run backward, but our lives and the natural world cannot. Similarly, an object or marker may move forward or backward relative to a "rigid" ruler or metrical grid; but the universe itself is not static, nor a mere product of definition.

Determinism holds that the motion of colliding balls on an idealized billiards table is theoretically calculable as far as one likes into future or past. A precise repetition of the initial conditions in such a system should give the same result.[26] A three-dimensional counterpart would be a collection of molecules moving in a closed container. But when the number of molecules increases to become a gas, one can no longer account for their individual motions and is led to a collective description. Then there appears to be an emergent tendency to reach an irreversible equilibrium, quite apart from reference to an environment.

How a theoretically reversible system becomes irreversible has been considered a serious problem in thermodynamics. It remains a mystery how the manifest irreversibility of the real world can emerge from the theoretically reversible motions of molecules. But this is something of a pseudo-problem, given that reversibility is a fiction to begin with and the motions of individual molecules cannot actually be tracked. There is a problem only after one has first forcibly constrained the molecules to a sub-volume of the container—a lower-entropy initial condition—from which to release the gas into a larger space. This larger volume serves as environment to the previous smaller one. The transition from reversible dynamics to irreversible thermodynamics thus depends on how the world is initially configured. This becomes especially important when applied to the universe as a whole, which has no environment. In cosmological terms, the mystery of the second law does not lie in irreversibility, but in how we conceive the initial entropy state of the universe.

Thermodynamics, which challenged the determinism of classical dynamics, arose from studies of reversible mechanical systems: idealized engines. Such an ideal mechanical system may be perfectly determinate and reversible by definition. In reality, however, there are no such systems. A gas sealed in a container, for example, might be well insulated so that there is no exchange of mass and very little of energy. But there is still an environment outside it, against which to measure change. There will still be a minimal energy exchange and no shielding from quantum tunneling or gravitational influence.

Equilibrium is a prerequisite for thermodynamics; temperature is even defined in terms of it. Yet equilibrium in the real world is exceptional. A system reaches equilibrium only in the absence of disturbing forces. Yet as we have seen, there are no such isolated systems in nature. Even if the universe as a whole is defined as a closed system, it is not clear that equilibrium would be a feasible initial condition or inevitable final state. The second law implies equilibrium as a final state, yet the second law itself circularly requires equilibrium.[27]

Identity and Individuality

The statistics of quantum measurements do not correspond to what one expects of classical particles, which can be distinguished one from another and assigned identity. The fact that elementary particles cannot be marked or tagged as individuals leads to a characteristically different statistical accounting for quantum entities. In quantum physics, there are difficulties of principle involved in finding the set of all objects (say, in a given volume) that satisfy the definition of a given particle type. In fact, in creating the statistics, it is never objects that are counted, but measurement events—which may represent quantities rather than objects. (Quantity can refer either to a countable number of individuals or to a number of arbitrary units into which a whole has been partitioned—for example, quanta of energy.) When quantity does not refer to individuals with distinguishable identity, it makes no more sense to speak of *this* unit as opposed to *that* one than it does to speak of *this* dollar as opposed to *that* dollar in a bank account.²⁸ This is hardly strange, since even in the classical realm energy had long been thought of as having quantity but not individuality.²⁹

As a modern-day Heraclitus might say, "You can never point to the same particle twice." The "river," like a bank account, is ever changing. Any background against which stable objects are to be identified is also ever changing. A macroscopic object is identifiable as an individual thing either by some permanent distinguishing feature or else by its unique space-time location in relation to other identifiable things or some imposed reference framework. Impenetrability is implied, for otherwise it could not uniquely occupy its position. In other words, constancy of the object and the background is presumed. But what does that entail? Can the object even be considered apart from relations to other things? How does one know it remains identifiably itself? To establish an identifying feature, an object must be compared to something—at least to some record (for example, in memory or a photo) or to another of its kind. The challenge of distinguishing two things of the same kind merges with the task of establishing the continuity of a single thing.

None of this strikes us as problematic for ordinary objects, whose individuality and genidentity we take for granted.³⁰ Upon returning to Paris, one is not surprised to find the Eiffel Tower where one left it, and not displaced to Las Vegas or interpenetrating the Arc de Triomphe. One accepts that the urban tissue surrounding it may change and that the structure itself may rust, be repainted, have flashing lights installed on it, etc. Yet the tower as a recognizable marker persists where it is. Though iconic, it remains one of a kind, distinct from other objects, in spite of numerous scaled reproductions.

But suppose that, long before the founding of Paris, an oak tree had marked the spot now occupied by the center of the tower's base. It would be far more challenging to distinguish this tree from other oaks surrounding it than to distinguish the tower from nearby buildings. While the oak tree is an example of a kind, it might still be recognizable as an individual, occupying precisely *that* spot. But how do we determine that spot without reference to an urban structure that does not yet exist, without surveyors or GPS, without reference to distant stars, which would depend on technology and a measurement grid? A reference frame of some sort is needed, which involves human intervention. The tree itself grows and changes with seasons, so may *not* retain permanent characteristics. It is one of a generation of trees in a background of forest; its persistence as an individual is temporary. At best, it will be replaced, approximately, by another tree succeeding it.

Protons, on the other hand, are believed to be very long-lived. They were once compared to billiard balls or ball bearings—mass-produced to be as identical as possible. (It would seem that God made particles identical, and this has become the industry standard!) The common denominator is that (unlike the tree) the billiard ball, the proton, and the Eiffel Tower are products of definition. In the case of the billiard ball, we have made something physical that corresponds to a definition; in that of the proton, we have made a definition to which something physical is supposed to correspond.

One may mark a billiard ball to label it as an individual, which changes only slightly its defining physical properties, such as mass. Alternatively, one may attempt to keep track of its position over time. This is as challenging as any shell game; it depends on a reference frame, on the ball's impenetrability (for otherwise it might occupy the same place as another object or be found under the same shell), and on the insignificant momentum of photons. While stratagems to establish individual identity may not work for protons, the point is that they only work for natural objects in general through an act of marking or of imposing some artificial context. A simple way to distinguish the tree in the forest is to mark it with an axe; a more complex way is to build a city around it.

An object is "classical" when there exists some way to unambiguously establish its individual identity. However, two billiard balls—or two planets—cannot easily be distinguished at a sufficiently great distance, and then perhaps only through side-effects of their motions on the light reaching us.[31] Though hugely macroscopic, very distant astronomical objects are often detected by microscopic events—a few photons interacting with electrons in charge-coupled devices or photographic emulsions. Unavoidable uncertainties thereby intervene in our knowledge of these putative bodies.[32] If such objects cannot be clearly distinguished by classical procedures, should they be considered to have individual identity any more than microscopic entities subject to similar uncertainties?

The individual identity of physical things rests on their discernibility. This in turn requires impenetrability, which depends on state. The identity of an ice cube, for example, depends on temperature. Hence, even the idea of an atom is relative to energy state, since its existence depends on forces that may be overcome—for example, in the degenerate matter of neutron stars or in the unification of forces at high energy.

At the quantum level there are only kinds, not individual things. Moreover, whether two things can even be held to be of the same kind—because they share all properties in common—depends on the possibility to enumerate those properties, which are taken to be definitional. Unless the world itself is a manufactured thing, real elementary particles are no more products of such definition than are the leaves of a tree. (They do not have a precise intension.) Clearly it is not *our* ability to distinguish and classify them that is essential to their functional relationship to other things. Just as the tree can specify where each leaf should grow, nature as a whole may be able to specify which elementary particle should be where and when, and under what circumstances particles are interchangeable. Though *we* cannot keep track of them, perhaps nature can.

Notes

1. Mandelbrot, Benoît. 2004. "A Theory of Roughness." A Talk with Benoît Mandelbrot. http://edge.org/3rd_culture/mandelbrot04/mandelbrot04_index.html.
2. Cf. Brigandt, Ingo. 2011. "Explanation in Biology: Reduction, Pluralism, and Explanatory Aims." *Science and Education* 22:69–91.
3. Jauch, J. M. 1973. *Are Quanta Real?* Bloomington: Indiana University Press, p. 64.
4. Barrow, John D. 1991. *Theories of Everything: The Quest for Ultimate Explanation.* New York: Fawcett/Balantine, p. 43.
5. Bell, John. 1990. "Against Measurement." *Physics World* 8:34.
6. Oreskes, Naomi, Kristin Shrader-Frechette, and Kenneth Belitz. 1994. "Verification, Validation, and Confirmation of Numerical Models in the Earth Sciences." *Science* 263:645 and footnote 25.
7. Yet even the problem of only three gravitating bodies has no easy solution.
8. Smolin, Lee. 2013. *Time Reborn.* Toronto: Alfred A. Knopf Canada, p. 215.
9. Barrow, *Theories of Everything*, p. 246.
10. Galileo. 1623. *Il Saggiatore* (The Assayer). He should have known better, since his own telescopic discovery of the mountains and irregularities of the moon demonstrated it was not a perfect sphere!
11. Cartwright, Nancy. 1983. *How the Laws of Nature Lie.* Oxford: Oxford University Press. See also Cartwright. 1999. *The Dappled World: A Study of the Boundaries of Science.* Cambridge, UK: Cambridge University Press.
12. e.g., Deutsch, David. "Physics, Philosophy, and Quantum Technology," pp. 2–3. http://gretl.ecn.wfu.edu/~cottrell/OPE/archive/0305/att-0257/02-deutsch.pdf.
13. Cf. Mets, Ave. 2012. "Measurement Theory, Nomological Machine and Measurement Uncertainties (in Classical Physics)." *Studia Philosophica Estonia* 5:171.

14. Wigner, Eugene. 1960. "The Unreasonable Effectiveness of Mathematics in the Natural Sciences." *Communications on Pure and Applied Mathematics.* 13:1–14.
15. Quoted in Holton, Gerald. 2000. *Einstein, History, and Other Passions.* Cambridge, MA: Harvard University Press, p. 64.
16. Davies, Paul. 2011. "On the Multiverse." FQXI Setting Time Aright Conference. www.youtube.com/watch?v=ulkulX8O5lI.
17. Ellis, Brian. 2002. *The Philosophy of Nature: A Guide to the New Essentialism.* Montreal: McGill/Queen's University Press, pp. 90–94.
18. Ellis, Brian. 2001. *Scientific Essentialism.* Cambridge, UK: Cambridge University Press, p. 214.
19. Cf. Callender, Craig. 1998. "The View from No-When." *British Journal for the Philosophy of Science.* 49:151–52. Final conditions could also be required.
20. Smolin, *Time Reborn*, pp. 117–18.
21. Lynds, Peter. "Time and Classical and Quantum Mechanics: Indeterminacy vs. Discontinuity." http://arxiv.org/ftp/physics/papers/0310/0310055.pdf.
22. For example, at absolute zero temperature, or zero density, or perhaps at the infinite density and temperature of the black hole singularity.
23. Cf. Stehle, Philip. 1994. *Order, Chaos, Order: The Transition from Classical to Quantum Physics.* Oxford: Oxford University Press, pp. 27–28.
24. As an algebraic property, such reversibility is called commutativity. Of course, a system may be defined to be noncommutative.
25. Cf. Ellis, George F. R. 2006. "Physics in the Real Universe: Time and Spacetime." http://arxiv.org/ftp/gr-qc/papers/0605/0605049.pdf: "The basic issue then is not why macroscopic situations involving chemistry and biology are time irreversible, it is why in some cases time reversible physics is a good approximation."
26. On a real table, of course, this is not possible. The slightest initial error is magnified exponentially as the number of collisions increases.
27. Equilibrium refers to situations where random fluctuations average out and there is no net flux of energy driving the system in any direction. This may not be the case in situations where the behavior of molecules is somehow coordinated. See Prigogine, Ilya, and Isabelle Stengers. 1984. *Order Out of Chaos: Man's New Dialogue with Nature.* New York: Bantam, pp. 180–81.
28. Paul Teller, Paul. 1998. "Quantum Mechanics and Haecceities." In *Interpreting Bodies: Classical and Quantum Objects in Modern Physics*, edited by Elena Castellani. Princeton, NJ: Princeton University Press.
29. Schrödinger, E. "What is an Elementary Particle?" In *Interpreting Bodies*, edited by Castellani, p. 207.
30. "Genidentity" is the continuity of something over time.
31. For example, remote planets are distinguished spectrographically through the Doppler effect, or by interpreting light curves of eclipsing planetary systems.
32. Quasars once fell in this category, hence called "quasi-stellar objects."

15

Is Nature Real?

> "We find almost no other reason for atheism than this notion of bodies having, as it were, a complete, absolute, and independent reality in themselves."
> —Isaac Newton[1]

> "It is the real world that keeps us honest."
> —Neil S. Turok[2]

Immanent Reality

Idealist and skeptical traditions have long held the world to be somehow illusory. Science and religion alike contend that the world revealed by the senses is not the fundamental reality. Medieval Christianity denied the immanent reality of nature by holding it to be a created artifact. The Enlightenment continued this tradition by discounting natural reality in favor of conceptual schemes. Ever since Descartes, the modern imagination has embraced the notion that the appearance of reality can be deliberately counterfeited. Computerization has granted new lease to that notion as recent generations have been entertained by virtual reality and other simulacra. Simulation has all but displaced reality in the popular imagination. The new byword is that the very basis of nature is "digital"—consisting fundamentally of "information." Physical reality itself is typically viewed as a computation.

The contrary view, summarized here, is that nature is real in a fundamentally irreducible sense. Certainly, what one normally *means* by reality is something that exists independently of human thought. This autonomy guarantees that nature cannot be fully mapped conceptually. It must transcend human thought, even while it *is* a human concept.

How we perceive or conceive reality is always a function of both subject and object. As a basic category, "realness" points not only to the world but also to its genetically engrained hold over the organism. In order to qualify as real, nature must be indefinitely complex; it cannot be a simulation, or an artifact with predefined structure, for this would render it a mere function of human thought. Autonomous, it resists formalization, containment in definition, and human control, defeating the reductionist program. It is characterized not by determinism but by randomness and irreversibility. *Change* is fundamental, not a timeless perspective imposed upon it. (Physics should now take inspiration from Heraclitus rather than Plato.) Finally, to see reality for what it is, the subject's contribution to the scientific view of nature must be fully acknowledged. Natural autonomy means that the object cannot be reduced to the terms of the subject. Natural philosophy evolved into first-order science by ignoring the significance of those terms. Future science must correct that error.

Propositional Knowledge

Anything made may be expressed in terms of propositional knowledge (for example, instructions on how to make it). Such an expression may *represent* something in the way that a *mental* image does but not in the way that an *optical* image does. A mental image may *seem* to be fleshed out with qualities like an optical image, so that they appear to be informationally equivalent. Yet they are not, for a mental image (like all made things) cannot be searched for more information than the propositional knowledge it already embodies.[3] Definite but limited information is encoded in any proposition, concept, mental image, program, simulation, text, or indeed in any other made thing. It is important to distinguish the information content of artifacts, finite by definition, from whatever structures in the real world they may be used to represent, whose information content is indefinite.

The question of the ultimate irreducible structure of nature cannot be separated from structures imposed by thought. Whether nature has a "bottom," or is indefinitely complex, is a question that is inextricably bound to how we think, but also to limits imposed on human knowledge and thought by physical reality itself—which, of course, includes the human brain. This interplay of subject and object leaves us with the challenge of how to distinguish a real bottom to nature's complexity from an apparent bottom imposed by thought. Such an imposition by itself would render physical reality ultimately a deductive system, a set of propositions. Dilemmas of the ancients regarding whether the world is a plenum or a void were later handled by the real number continuum, which provided a rigorous treatment of geometric space and was the basis of the physics of differentiable curves and equations.

Yet, paradoxically, this treatment of continuity is defined in terms of limits of a series of propositions.

Natural Resistance

The pre-Socratic natural philosophers had set out to rid the world of gods and superstitions, of plan and purpose supernaturally imposed. Things simply follow their inherent natures, and whatever order exists is intrinsic to the cosmos, not imposed upon it by some external agent. Aristotle embraced this view, in tune with the archaic notion of the immanent reality of nature. It was at odds, however, with the Platonic and Pythagorean philosophies, which insisted that reality must be the product of a rational mind, whose order is forcibly imposed upon a chaotic substratum. From that perspective, the resistance of physical reality to the impositions of thought renders the natural world imperfect, if not pernicious and evil. This idea appealed to the Christian mind, along with divine creativity externally imposed.

The exclusive interest in efficient cause in the early modern period affirmed that matter only passively suffers action upon it. An effect had been thought to last only as long as its cause, so that ongoing motion required an ongoing motive force to keep it going.[4] In the absence of a dynamic concept of mass, inertia was not viewed as an intrinsic property of matter, but more like motion through a viscous medium.[5] Yet, mass seems to express the material reality of things—their very essence as distinguished from other sensible qualities. The fact that it cannot be defined apart from such qualities (if only change of position) has rendered it an elusive concept in physics. In the *Principia*, finally, Newton gets it right: inertia is a "power of resisting, by which every body... continues in its present state, whether it be of rest or of moving uniformly forwards in a right line." Matter, then, was not completely powerless. While unable to move itself, at least it did have within itself the ability to resist being moved by an external force.

In Plato's view, the source of order and perfection is the mind, while the body—and matter generally—is imperfect, unruly, subject to decay and prone to disorder. Plato is correct in his assessment that the "imperfect" physical world frustrates the inclination to view it as simple or subservient to human purpose. Possessing their own particular properties, real materials resist human manipulation, as any artist or engineer knows. An industrial product, for instance, is but an imperfect realization of the design—requiring, moreover, enormous waste. Not an industrial product, the very nature of nature is to resist tidy schemes that attempt, in effect, to treat it as though it were. The fact that the world is complex and messy is a sign that it is real rather than ideal. Reality lies precisely in such "imperfection." As well as the resistance of inertia, nature's resistance to simplifying thought (along with the absence of waste) is an essential aspect of its autonomy.[6]

"Realism"

Realism, in the scientific context, generally presumes a view of inanimate matter as passively obedient to externally imposed laws, not imbued with its own active powers or principles of self-organization. This view had been handed down from antiquity through Christian theology. As Philo of Alexandria had put it: "To act is the property of God, and this we may not assign to any created thing; the property of the created is to suffer."[7]

On the other hand, one may argue that the Enlightenment notion of divine whim supports the contingency—and therefore autonomous reality—of nature. For nature and God are alike beyond human ken. The unpredictability of divine action opposes itself to the idea that the universe must be the way it is by logical necessity, which presumably even divine will could not defy. Yet one must consider also that the universe might be just as it is for reasons other than *either* logical necessity *or* divine intervention. After all, our planet is a self-organizing and self-sustaining ecosystem that cannot readily be derived from first principles; it results from the complex interaction of many factors, which theory may never fully map.

A paradox of realism is illustrated by the inconsistent development of the concept of mass, which had to fulfill a double role: as the self-existing essence of matter and as a measurable quantity. On the one hand, mass stood for the very substance of things, independent of human intervention. On the other, to be measured it depended on the application of a force. The contradiction between these roles is reflected in Newton's mutually defining concepts of force and mass.[8]

While realism often means that properties exist independent of the observer, obviously no knowledge of these properties can be independent of observation or measurement. Aside from observer-independence, however, another symptom of nature's reality is its essential resistance to formalization.

Realism is generally at home in classical physics, where imprecision was not taken as indeterminism, much less to mean that the world has no definite existence or properties. Rather, the fact that the *mathematics* worked precisely—within the limits of imprecise measurement—was taken to mean that some real *causality* worked perfectly behind the scenes. This amounted to the assumption that physical variables must *have* precise values, even when they can only be measured imperfectly.[9] But "physical variables" are idealized constructs, and the equations in which they appear are precise by fiat. Their identity with physical reality is merely presumed, based on an approximate correspondence with measured values.

Realism is less at home in the quantum realm. According to the Copenhagen interpretation, it is meaningless to ask what state a quantum system is in prior to measurement. One takes for granted that ordinary objects continue to exist and have measurable properties between sightings, because of the common experience of re-locating them and measuring their state, which is then seen to have continuity with previous measurements. However, quantum objects are not identifiable in that way.

In realist terms, Schrödinger's cat[10] is either in the box or not, either alive or dead, quite apart from anyone looking to check. The Copenhagen interpretation is often construed to mean that the cat is in a "mixed state" before the box is opened. The state of the cat is supposed to correspond to the wave function of a whole system that includes a decaying atom and a cat, whose individual wave functions are superposed. Viewed this way, one must conclude, with Einstein, that such a *single* system is not "completely" described by the quantum theory, which cannot predict the state of the cat. Cruel as it would be, a more fitting analogy would be an *ensemble* of such systems—many trials of cats in boxes—which would approach a statistical average: alive half of the time and dead half of the time. It is absurd to speak of one cat in one box as being in limbo, between dead and alive, when it is simply our knowledge that is in limbo.

The debate between Einstein the realist and Bohr the positivist reflects the general philosophical question of whether physics represents nature or our knowledge of nature. A description can be complete in regard to the state of knowledge, while incomplete as a description of reality. In that sense, Bohr and Einstein were talking at cross-purposes. A probabilistic description is necessarily incomplete from a deductionist perspective. The state of the box (before opening) is understood differently in the two perspectives. Einstein expressed common sense in pointing out that there can be no intermediate state between an alive and a dead cat, between an exploded and an unexploded bomb.

Yet even if we cannot pinpoint what causes a given bomb to explode or not, there can be a statistics of how many bombs fail to explode in the course of many trials. For Einstein, a complete theory is one that maps a (single) system on a one-to-one basis; but this is not feasible unless that system is a deductive system.

Simulation

Unlike man-made things, the reality of nature resides, so to speak, more in the "detailing" than in the "blueprints," precisely because it is the details that elude definition and mathematical formulation. Now, a simulation can provide both large-scale order and seemingly random detail. What distinguishes simulation from reality, however, is that an algorithm *generates* the detail in the simulation. In contrast, one only guesses at an algorithm to fit the observed patterns and details of nature after the fact.

A law of nature, when mathematically expressed, algorithmically compresses observed data.[11] Unless nature itself is nothing but a data set that has been generated by the algorithm in the first place, the algorithm only approximately fits the natural reality. In contrast, it may *precisely define* a simulation. Moreover, the immanent reality of nature implies that laws of nature are not transcendent, existing independently of matter; they are neither immutable nor governing. Not only the details but the laws too must be immanent in matter, not imposed on it externally. The apparent universality and constancy of physical laws is an empirical finding, not a necessary or transcendent truth.

If nature were literally a simulation or artifact, generated by an algorithm, it could not manifest indefinitely varied or random detail. In that instance, moreover, no physical quantities would be truly random numbers. Conversely, *if* it could be shown that a natural constant is truly random, this would prove that the natural world is not a simulation or artifact. Unfortunately, while there is no way to mathematically generate true randomness, neither is there any way to prove it. On the other hand, a sequence can be shown to be *not* random, simply by finding an algorithm that generates it. Thus, *if* nature is real, there is no way to exhaustively simulate it; on the other hand, there is no way to prove that it *is* real—any more than one can prove the existence of God. In the absence of absolute proof, there are nevertheless more plausible arguments for the immanent reality of nature than for the existence of God. We have mentioned nature's general untidiness and quantum weirdness as examples.

Mechanism, insofar as it implies an imposition from the top down, is the very thing that *cannot* explain physical reality as a self-organizing process from the bottom up. Like other machines, a computer running a simulation must be set initially in motion. Though it consists of commands in a "language," the language has no meaning to the computer itself. Whether applied to the cosmos or to the brain, the computer metaphor continues to ignore the most obvious fact of physical reality, which is that it is *not* programmed by some external intelligence, but is somehow self-programming and self-activating.

A Digital Universe?

Digital physics is associated with the notion that physical reality is nothing *other* than information, mathematics, simulation, computation, or geometry.[12] This disposition is reflected in various modern proposals, such as the mathematical universe hypothesis;[13] John Wheeler's "it from bit" philosophy; diverse schemes of posthumanists; and in a variety of approaches that can loosely (if awkwardly) be labeled "ontic pancomputationalism." Though mathematics effectively describes and anticipates many features of the world, any assertion that physical reality is *reducible* to mathematics, let alone to digital computation, is purely metaphysical. While dressed up in the latest technology, such notions hark back to the idealism of Plato and Pythagoras. It is no more plausible, however, that physical reality should correspond to the computers and software of the twenty-first century than that it should correspond to the clocks and waterworks of the eighteenth century or the steam engines and mills of the nineteenth. Of course, we are saturated with information technology and fascinated by it, as these earlier times were with their mechanisms. While it is understandable that computation is the contemporary metaphor, all metaphors must be taken with a grain of salt.

The "Simulation Argument"

Similar to the Doomsday argument,[14] the simulation argument reasons (absurdly, I believe) that we are probably living in a computer simulation.[15] Such a conclusion rests on dubious assumptions, most obviously that there can even be such a thing as "living in a simulation." While suggested by the technology of virtual reality, this notion stems from the skeptical tradition introduced by Descartes. Virtual reality, however, presumes a real subject living in the real world. The subject undergoing experience of a simulation is a biological creature, and the simulation interfaces with the subject's real sensory apparatus. This is an entirely different concept from that of a *simulated subject*, which exists entirely as lines of code in a computer program. It is a further assumption that this virtual subject (with virtual brain and body?) could be conscious and have experience. A virtual reality, as we currently understand it, is an entertainment created for the enjoyment of real human subjects. As such, no part of a simulation could *be* a subject any more than a fictional character in a play can be. This is not to deny that artificial subjects could exist—provided they had real (though artificial) bodies, which evolved under the sort of conditions that shape the emergence of natural subjects.

It is commonly held that "mind" is an abstraction that does not depend on a particular physical infrastructure and could "run" (like software) on a silicon-based physiology, for example. Some conclude from this that a computer could be conscious and have experience. However, the particular substrate is not the issue. A computer is not a brain—not because it is made of silicon, but because it does not have a body for which it evolved as an organ of survival. Computers as we know them are not *embodied*. They do not, like organisms, have their own priorities, established through an evolutionary history. The idea that a string of code, or a sophisticated piece of hardware, could be programmed externally to have its own priorities is self-contradictory.

Of course, one could argue that *nature* has programmed the organism, and that human programmers (or advanced alien programmers) should be able do something analogous with computers. However, as far as we now know, nature does not operate from the top down, as though there were a programmer at work on high. Such reasoning assumes some duality within nature corresponding to the relationship between programmer and computer, mind and matter, creator and created. While it seems plausible that artificial organisms could evolve toward true intelligence, under the right circumstances, that scenario is (mercifully) a far cry from the present disembodied computer. And even if computers, as embodied artificial organisms, could evolve consciousness, what would it mean for a *simulation*, running within such a computer, to be conscious? The situation bears comparison to dreaming: one can imagine a character in one's dream to be conscious, yet that impression is one's own mind at work, not the mind of the dream character.

Like the "brain in a vat," the so-called simulation argument updates Descartes' paranoid fantasy that information can be fed into one's brain from a source other than the senses. The simulation argument adds a touch suitable to the modern taste: *given* advanced civilizations running simulations in which simulacra *can* be conscious, what is the probability that you are one of these simulated creatures and that the world you find yourself living in is a simulated world? The reasoning of the argument goes like this: Just as there are far more programs than programmers, there must be far more simulated than real subjects. Ergo, given just the numbers, you are probably a simulation yourself! Such an absurdity has no meaning unless one assumes, in the first place, the very thing that is to be concluded: that "reality" is not real.

Notes

1. Hall, Rupert, and Marie Boas, eds. 1962. *Unpublished Scientific Papers of Isaac Newton.* Cambridge, UK: Cambridge University Press, p. 144. (The editors place the writing no later than 1672.)
2. Turok, Neil S. 2012. *The Universe Within: from Quantum to Cosmos.* Toronto: Anansi, p. 14.
3. Sartre identified the key difference between the real and the virtual (in his terms, between sensory and eidetic images), namely, *detail*. Limited information is encoded in a mental image compared to a sensory image.
4. Jammer, Max. 1997. *Concepts of Mass in Classical and Modern Physics.* New York: Dover, p. 71.
5. Ibid, p. 20. This bears an ironic resemblance to modern theories of the origin of inertial mass through interaction with the Higgs field, which suggests material properties of space itself.
6. The "law of requisite variety" reflects the fact that the model can never be as complex as the reality it models. The conventional division between laws and initial conditions may reflect the inevitable remainder, outside theory, of details that escape the rule. Finally, irreversible processes may be viewed as nature's resistance to experimental control See Prigogine, I., and I. Stengers. 1984. *Order Out of Chaos.* New York: Bantam, p. 120.
7. De Cherubim 77.
8. Jammer, *Concepts of Mass in Classical and Modern Physics*, p. 89ff.
9. Powers, Jonathan. 1982. *Philosophy and the New Physics.* London: Methuen, p. 140.
10. A thought experiment to explore the role of amplification between the microscopic and macroscopic levels. A cat is enclosed in a box with a radioactive substance, which—unpredictably—can emit a particle that will trigger a poison gas to be released in the box, killing the cat.
11. That is, it expresses the data set in a form shorter than the data set itself.
12. As for geometry, see for example: Lisi, A. Garrett, and James O. Weatherall. 2010. "Geometric Theory of Everything." *Scientific American* 23:96–103.
13. Tegmark, Max. 2007. "The Mathematical Universe." http://arxiv.org/abs/0704.0646.

14. First proposed by astrophysicist Brandon Carter in 1983, the Doomsday argument reasons that, supposing the present occupies a random place in the human timeline, then we are probably halfway through it.
15. Bostrom, Nick. 2003. "Are You Living in a Computer Simulation?" *Philosophical Quarterly* 53:243–55. Bostrum himself only states that this is one of three possible conclusions based on extrapolations of current technological trends. The other two are that most civilizations become extinct before achieving that capability or that they are just not interested in pursuing it.

Part Four

Beyond the Mechanist Faith

16

Is Reality Exhaustible in Thought?

"Reality was but a creature formed from one of the intellect's ribs, from language. We could take care of her, fill her up, and leave her spent."
—Rebecca Goldstein[1]

"The most important thing about the natural world is that it is what it is and not something else."
—Roberto M. Unger and Lee Smolin[2]

Formal Representation

To what extent is it possible to reverse-engineer nature? To some, physics is the quest for a completed mathematical description of all the basic forces and entities in a few equations, leaving no further major surprises. At the very least, such optimism assumes that nature is finitely describable. I have argued against this, since it would mean that nature itself is nothing more than the description. We have seen that such a notion, however nonsensical, has deep and widely unacknowledged roots in our religious and idealist heritages. This did not prevent some early critics, such as Kant, from wondering how a truly empirical, contingent, or synthetic natural science could lend itself to purely formal treatment within deductive systems.

Nevertheless, a neo-idealist view is that the natural world lends itself to mathematical representation because it literally *is* mathematics. It is not surprising that mathematicians would hold such a view; it is more surprising when physicists do. One thinker, who wears both hats, declares that "our successful theories are not mathematics approximating physics, but mathematics approximating mathematics!"[3] However, to treat the universe not just as mathematically *describable* but as mathematics *tout court* is simply to reduce matter to mind.

To be mathematically described at all, a natural entity or process must first be formally redefined. It is this idealization that is then the object of theoretical, experimental, and mathematical treatment. A natural entity, however, cannot be fully specified or exhaustively described without referring to broader contexts with which it is indefinitely interconnected. Short of such inclusion, it is a product of definition, and not the natural entity presumed. While such an artifact *can* be exhaustively described using the tools of propositional thought (including digital computation), there is no guarantee that it corresponds perfectly to the natural entity or process it concerns. The laws of nature are such artifacts—simple expressions of patterns teased out from more nuanced appearances—often then recycled as the cause of what they describe. They are very useful in situations that correspond more or less to an ideal—for instance, the influence between two gravitating bodies such as the sun and a planet. However, by definition they may exclude more subtle factors perhaps not yet discovered.

The faith behind the notion of a complete theory, a perfect simulation, or an isomorphic model is that each and every property or element of a thing can be formally represented. This is the classical ideal of a one-to-one correspondence between the elements of a physical theory and those of the physical reality it purports to describe. However, since the elements of the natural world cannot be determined unambiguously, nor exhaustively, such a complete specification is possible only in a deductive system, not in nature. The only possible complete map of the world is the world itself.

To exhaustively model a real system is equivalent to finding an algorithm to express a truly random sequence—a contradiction in terms.[4] An algorithm can only generate or compress a pseudo-random sequence.[5] There is a limit to the ability of propositional thought to formalize reality, and nature is "real" precisely because it cannot be exhaustively formalized in any theory or computation. No theory, therefore, can be complete in this sense. The probabilistic nature of nature is the very sign of nature's irreducible autonomy. Determinism, on the other hand, expresses the wishful thought that nature can be accounted for exhaustively from first principles, in an axiomatic system.

A theory could be complete in the mathematical sense, as a self-contained deductive system, but this would only limit its power to represent nature.

Is Necessity Necessary?

Following Hume, causal necessity exerted by one thing or event over another—such as we imagine "forces" to involve—is not a coherent notion. One speaks of *cause* when events are correlated in a way that appears to satisfy logical conditionals (necessary and/or sufficient conditions). This requirement enshrines the need for certainty regarding the future, while reflecting only the past. The only true necessity is logical necessity, which is always a matter of definitions rather than empirical facts.

The notion of causal connection serves to rationalize observed correlations and justifies expectations to the extent that future events are correctly predicted. But the appearance of necessity itself results from conflating observed correlation with logical implication. It serves a psychological purpose but bears no stamp of guarantee: the pattern may or may not hold in the particular instance or in the future. It facilitates certainty in such cases, when in fact there are only patterns observed among past events or estimates of probability based on them after the fact. We are reassured to think that physical reality can be redefined in such a way as to yield analytic truths when we are only confronted with empirical evidence. We naturally prefer to believe the world corresponds to our definitions and wishes. Successful mathematical treatment of nature (especially prediction) majorly confirms that faith but at the cost of focusing only on those features of reality that exhibit clear patterns that can be expressed mathematically. We choose to look at simplified aspects of nature because these are what we can master technically. Because they are effectively mechanical, mathematical procedures and models can lead to the construction of real machines (which are *designed* to follow patterns); and they can lead to physical manipulation of nature that takes advantage of its apparent patterns. As computation refines and extends our concepts of mechanism, and expands what is technically feasible, the scientific focus on the type of systems studied can and should expand accordingly. The study of complexity is already an example of this broadening.

We have seen that determinism is an ambiguous concept. It refers, on the one hand, to what one can *discern* and, on the other, to a putative *power* of one thing over another. As Hume argued, this power is fictive, at best based on the human experience of making things happen. Since there is no natural determinism, there is no natural indeterminism either. Just as the only determinism is logical necessity, the only indeterminism is logical undecidability. Like determinism, "indeterminacy" is not a property inhering in nature, but a state of knowledge.

Properties, Parts, and Propositions

The identity of an individual thing is relative to the parts that constitute it as a whole, and also to the greater wholes of which it is a part. Moreover, like the partitioning into "system" and "environment," a property or element reflects a human cognitive act as well as a putative reality. It involves an assertion that disregards indefinitely many other assertions, following specific purposes. The collected properties or parts of a natural thing do not (as assumed in reductionism) constitute the thing as a whole, although they *do* constitute an artifact. Such an artifact is exactly and only what it is defined to be. In contrast, nature is ambiguous and does not correspond perfectly to our definitions. In general, one can reversibly take apart and reassemble artifacts, which have clearly defined parts. This is not usually possible with natural things without serious consequence, since they do not come with well-defined parts.

Any list of properties that thought could assign to a natural phenomenon is limited. It cannot exhaust the being of a natural thing, which has indefinite properties. However, it can exhaustively describe an artifact; more to the point, it *defines* an artifact. While we cannot know all there is to know about Shakespeare as a real historical figure, we *can* know all that there is to know about Juliet, her fictional balcony, or her thoughts about Romeo. In principle, we could discover more about Shakespeare (and even more about a living author)—but not about Juliet. Similarly, we cannot exhaustively know a natural reality, although we can exhaustively describe a theory of physics and list its elements and propositions.

When from the outset a natural thing is *taken* to be a collection of properties or parts, it is mistakenly assumed to be exhaustible. This is the fallacy of the mechanist, reductionist, or deductionist stance. Once the circularity of this reasoning slips by, one predictably may fail to see any difference between reality and models or simulations of it. It is then tempting even to see the whole world as virtual and one's own existence as nothing but numbers being crunched in a vast computation. This is simply a technologically updated version of Descartes' paranoid thought experiment in which the "evil genius" is replaced by a computer to counterfeit one's sensory input. The answer to both is the same: in principle, there *is* a way to tell the difference between reality and simulation, for no simulation can perfectly match the richness of detail that characterizes natural reality. A crucial difference between simulation and reality, as between digital and analog, is that the former contains limited propositional knowledge, and therefore limited detail. Simulation is intensionally well defined, but extensionally sparse. Reality, in contrast, is ambiguous but dense.

Scientific modeling is essentially simulation. Mathematical equations define and describe formalisms that could be expressed as computer programs. A theory, equation, or program corresponds only selectively with natural reality, touching it at points, as the integers touch the real number continuum at literal points. But other aspects of reality may lie outside the simulation, just as there are real numbers that are not integers. Equations and the models they describe are isomorphic to each other, because they express the same formalism. But no formalism is strictly isomorphic to the real process it models, corresponding only in specific and limited ways.

Models

Science represents the world by means of models, whose *logical* relations map observable *natural* relations. However, the equivalence is always only approximate. While the relationships defined in the model are necessary and precise by definition, those in nature may not be. While models may be isomorphic to each other, this says nothing of their relationship to reality.

The flight of a model airplane is analogous to that of real aircraft in a literal sense because it *is* a real aircraft reduced in scale. On the other hand, human flight mimics the flight of birds only in an abstract or metaphorical sense. A model bird or insect, even should it fly, is not a real bird or insect but an artifact, like the full-sized aircraft or its toy version. The idea behind the universal machine (digital computer) is that it can simulate any other *machine* exhaustively. But whether a machine, program, artifact, or any other intentional construct can exhaustively simulate an organism—or *any aspect of natural reality*—is quite another question.

The use of linear equations in physics is limited by the fact that natural phenomena generally involve *non*-linear processes. Yet if nature is *indefinitely* complex, there is no hope to capture it in *any* algorithm whatever, not even those that generate endless detail, as in the Mandelbrot set. There is no guarantee that any mathematically generated pattern corresponds to natural reality. There is only the trivial guarantee that an algorithm corresponds to the pattern it mathematically generates.

A bottomless nature poses a dilemma for fundamental physics, since it resists ultimate rational ordering.[6] The laws of physics are algorithms, whose predictive value depends on computable numbers. If nature is indefinitely detailed—corresponding to the real number continuum—those aspects of it accessible to calculation are infinitely outnumbered by aspects that correspond to non-computable numbers.[7]

The complexity of nature is both revealed and hidden by our ways of looking. Surrounding and permeating us, the universe no more readily appears to us "complex" than water does to the fish. We have learned, of course, to discern complexity through analytical eyes, which have been trained to see parts and the relationships between them. But such are artifacts of our thought and seeing, often modeled after the design principles of made objects, including axiomatic systems. The complexity we see may resemble that of our artifacts, insofar as we see to the depth we have been able to create—which may be far less than the real complexity of nature.

Chunking

Perhaps what makes the prospect of perfect simulation seductive is the characteristic "chunking" involved in language and thought alike, whereby "a rose is a rose is a rose." Yet there are many varieties of rose and every individual flower is unique. In particular, an artificial flower is not a real one. A baseball player and a pitching machine may alike be called *pitchers*, but the machine only crudely and metaphorically imitates the human. The flight of the aircraft is only metaphorically the flight of the bird. When a "piece" of behavior seems to resemble another action, it is not being compared to the action itself but to a common definition, which is held to underlie them both and which has been abstracted as the essence of that behavior or action. Similarly, one may falsely believe the essence of a natural object or process to be captured in a program or blueprint for its artificial reproduction. This is the delusion on which conventional top-down artificial intelligence has foundered.

A truly convincing digital simulation is computationally costly (which argues, perhaps, against a simulated universe!) Even so, the useful approximations afforded by computation may blind us to the fact that any model or simulation *cannot* be exhaustive and is also *qualitatively* different from the natural thing it models. Common sense recalls the differences between real and artificial things, or between things and their representations. While artificial flowers are intended to deceive, no one should confuse the intricate muscular action of *throwing* with the mechanical hurling of the ball. Familiarity and practicality, however, may have bred neglect of the real differences between the flight of the bird and that of the aircraft. The very concept of simulation rests on obscuring such distinctions, by conflating all that can pass semantically under a given rubric. The algorithm, program, or formalism is the bottleneck through which the whole being of the real phenomenon must pass in order to be simulated.

Simulation is not duplication, except where *artifacts* are concerned. One thing simulates (and duplicates) another only when they embody a common formalism—as with two computer files, for example, or the airplane and its model, which are but alternative constructions from the same design. The being or behavior of a natural object or phenomenon, however, is not exhausted in the program, formalism, model, or ideal abstracted from it. It is a mistake to hold that this formalism can serve as a blueprint for its reproduction. As Leonardo attempted, it might be feasible to build a mechanism that flies like a bird rather than like an airplane. Even so it is not a real bird. The natural thing is a found object, not an invention constructed from design. It does not come with a blueprint, which can only be imposed after the fact through a structural analysis that can never be guaranteed complete. The mechanist faith involved in some aspects of molecular biology, nanotechnology, and artificial intelligence proposes that it is possible to duplicate or reengineer natural systems by reverse-engineering them first. This is a fallacy, however, for the structure and behavior of the natural thing cannot be exhaustively formalized. The artifact *will* instantiate the formalism, of course. But it will *not* duplicate the natural object, any more than an airplane duplicates a bird.

Turing Tests and Generative Codes

A computer program (animating an android, for instance) might simulate human or animal behavior in a detailed way, even down to the level of involuntary physiological response. The tacit premise is that all aspects of an organism's behavior as a physical system can be reduced to the equivalent of a written message. While this is merely an article of faith, it is the working hypothesis of artificial intelligence. The heuristic of the Turing test—that if we can't tell the difference then there *is* no difference—becomes a technological fudge factor, in which subtle differences and their potential consequences are ignored for practical purposes. Molar behavior is imitated, without thorough regard for the pathways leading to it. Subtle differences may ultimately be crucial for theoretical purposes and also for our view of reality. Ignoring them may lead to absurd expectations, such as the pipe dream of "uploading" human minds into computers, where they can supposedly live forever in cyberspace or be "downloaded" into cloned or artificial bodies.

Computer simulation rests on the principle that everything can be approximated with arbitrary precision by a digital program. This is a useful principle for entertainment devices, whose intended product is not a truly exhaustive replication of an original but a subjectively satisfying simulacrum. In other words, simulation is useful to deceive eyes and ears that do not need or want to tell the difference. It is incapable, however, of duplicating or exhaustively modeling reality.

Simulation is useful in many ways, both in scientific research and in education. Before the computer revolution, the depiction of an imagined perspective—such as the view from a space probe approaching another planet—was presented in books and films as an "artist's conception." Computer-generated imagery (CGI) has become so sophisticated that such depictions are now routinely integrated into real footage without disclaimer. (Actual photos from the Hubble telescope, for example, are integrated with CGI depictions of black holes, quasars, etc.) Perhaps the deception is innocuous for educational purposes. It is rationalized by assuming the verisimilitude of the computer images and the validity of the speculations behind them. Experts may recognize the difference but the lay public, for whom they are often intended, are not in a position to know what represents empirical data and what simply depicts current theory.

Critics of contemporary genetic science see in it a new form of the old mechanism. A tree's genetic "code," for instance, may in some sense be a set of instructions on how to grow the tree. But they are instructions to the natural world, inside and outside the cell, not to human engineers aspiring to play God. Hacking the tree's "program" must be implemented by the ability also to "operate" the natural environment— to secure and appropriately manipulate the resources needed to build the tree from scratch, enlisting the cooperation of other organisms, etc., exactly as nature does. The notion that the DNA of the tree contains all that is needed to unfold the tree's development, just as a program computes from its instructions, harks back to the misogyny of Aristotle, for whom the creative principle lay exclusively within the male seed. It denies the role of environment (even the "soil" of the womb), as well as the role of internal processes (such as may involve "junk" DNA) that happen to be extraneous to the program as defined. Just as we cannot credit the organism's development exclusively to a genetic code, so we cannot credit the unfolding of the universe exclusively to a "cosmic code."

Notes

1. Goldstein, Rebecca. 1983. *The Mind-Body Problem*. New York: Random House.
2. Unger, Roberto M., and Smolin, Lee. 2015. *The Singular Universe and the Reality of Time: A Proposal in Natural Philosophy*. Cambridge, UK: Cambridge University Press, p. xi.
3. Tegmark, Max. 1998. "Is the 'Theory of Everything' Merely the Ultimate Ensemble Theory?" *Annals of Physics*, 270:1–51.
4. Cf. the "Halting Problem," which, like Gödel's theorems, can be interpreted to describe the relation between formalisms and the real systems they model.
5. i.e., one that appears to be random but is actually generated by an algorithm.
6. Barrow, John D. 1991. *Theories of Everything: The Quest for Ultimate Explanation*. New York: Fawcett/Balantine, p. 103.
7. Hartle, James B. 1997. *Sources of Predictability*. http://arxiv.org/abs/gr-qc/9701027. While there are countably infinite possible theories, the real numbers are *uncountably* infinite. If nature corresponds to the real continuum, it is infinitely more complex than any theory, even an infinite number of theories!

17

Mechanism and Organism

> "The question in the end is whether the world is
> more like a dodecahedron or more like a flower."
> —Lee Smolin[1]

> "Anyone can make things, if they will take time and trouble enough,
> but it is not everyone who, like me, can make things make themselves."
> —Charles Kingsley[2]

Introduction

Some things appear to be self-animating, others to be passively subject to what we now call laws and forces. The philosophy of mechanism has been very successful in the study of the latter kind of things, especially for creating technology. It is now applied to living things as well, in genetic technology. As they become ever more sophisticated, there is some irony in the fact that machines are increasingly designed to resemble organisms. Nevertheless, mechanism remains the reigning paradigm, especially in the guise of the computational metaphor. An underlying metaphysical premise tacitly remains: nature is equivalent to an artifact. To qualify as a scientific hypothesis, this would have to be experimentally testable. Yet to serve as an approach to nature, with implications for research strategies, requires no more than acceptance.

Whether the world is better viewed through the eyes of mechanism or through a more organic metaphor is tied to the question of the meaning of physical law. *Mechanism* views law as fundamental and transcendent, a creative force separate from the universe it rules. Laws are the principles according to which a machine is designed to operate. In contrast, *organism* views physical law as description of collective behavior that is immanent in matter. We are now discovering that the macroscopic properties of matter cannot simply be reduced to effects of Newton's laws on the microscopic scale. Rather, Newton's laws may be seen to emerge on the macroscopic scale as effects of the collective organization of matter.[3]

The study of the very large and the very small each fostered the quest for simple first principles. In contrast, the intermediate human scale is acknowledged as more complex. This is partly because this scale is dominated by organic phenomena, but also because it is more readily accessible to direct investigation. It is easier to interact with things close to our own size, which makes it easier to perceive multiple causes and to trace complex interactions without the need for oversimplification. Suppose, however, that complexity actually prevails on *every* scale, yet we have missed it because of an obsession with simplicity? Suppose that "inert" matter has unrecognized abilities to self-organize, making the cosmos more like an organism than a machine?

The gains of the mechanist approach come at a cost. For the mechanist, causation means efficient cause. If matter is "mechanical," cause can involve only a passive transfer of energy, motion, or influence from one entity to another (as reflected in the laws of conservation of energy and momentum), but never its origin. Such a system is passive as a whole, as well as in its parts; it lacks the ability to set itself in action. While the power of nature to actively self-organize is crucial to its autonomy, it is only by ignoring this autonomy that nature has been so successfully treated mathematically. Yet this very success has led to the impasse that mechanism cannot account for the properties of the whole or its beginning. If nature is *not* such an inert machine, then it must retain within itself the ability to set itself in motion. How this capacity is distributed or localized within nature, and how it functions, might be compared to the localization, distribution, and (self-)organization of function within the brain. And just as brain function has only relatively recently begun to be effectively studied, so the self-organization of matter in general terms is but a recent field of study.

In Aristotle's static world, the question of self-organization in a modern sense could not arise. In Reformation thought, the notion of God's radical sovereignty reserved creative power exclusively to God, which the philosophy of mechanism supported by limiting the power of nature. Kepler imagined gravitation as a kind of spiritual force emanating from the sun—harking back to solar religion and vaguely implying a vitalistic cosmos.[4] Newton believed that non-living matter does not contain within itself the capability to move or otherwise change: all change must be externally imposed. "Force" was embraced as an impersonal agency, though it had once been conceived in personalized terms.

Yet what can "external" mean, in terms of the cosmos as a whole? In the Newtonian worldview, a force acting upon the cosmos as a whole would have to come either from some region outside it—which appears to be a contradiction—or else from some entirely *non*physical source. Short of divine intervention (Newton's answer), reason requires that, in some sense, the cosmos must activate itself. Macroscopic things give the false impression of passively following simple mechanistic laws of motion, an impression that is only feasible through the convention of isolated systems. More accurately, every thing must influence, and be influenced by, every other material thing. This interconnectedness makes the vision of organism far more plausible.

First Cause

Unless a system is conceived to be *self*-activating, change within it can only be caused by an agent outside it. We have a strong propensity to conceive such an agent in human terms; an "inert" system that self-activates seems to us a contradiction in terms. In other words, we continue to think in terms of a duality of mind and matter.

Of Aristotle's four types of cause, only "efficient" cause survives into the modern era. According to this notion, one system may serve as a cause of processes happening within another. A causal chain is effectively a series of events transmitted through a suite of defined subsystems, each of which serves as external cause for the next. The problem is that there is nowhere for the buck to stop. It requires an unending series backward in time—even to a time before "time" could meaningfully be defined. Alternatively, it requires a "first cause" that permits no further analysis because it is an essentially different kind of event. Psychologically, the model for such a cause is human will (even when idealized and extrapolated as divine will). We learn in infancy to cause our own limbs, among other things, to move. It is a later childhood achievement to project this power upon mutual interactions among "inanimate" objects so that one thing can serve as external cause for changes in another.

In modern times, the concept of agency, even as it pertains to human beings, is explained away in terms of impersonal interactions, ironically reversing our early-childhood understanding. The advance of science has encroached on theological or teleological accounts, reducing agency to efficient causation. We have witnessed a continual expansion of the limits of what can be explained mechanistically, with a corresponding diminution of the role of personal agency. Yet there remains always something outside the causal system, which either remains unexplained or must be accounted for by a first cause. There is still room for a "God of the gaps." However, the problem of first cause is not primarily theological but logical. It is a scientific problem insofar as science is committed to logic as we conceive it.

Perhaps logic itself must be revised. Like mathematics, it reflects general expectations regarding the behavior of things, general experience on the everyday human scale. Nature, however, defies commonsense expectations in unfamiliar realms. It was a sweeping revelation of Newtonian science that things appear to behave similarly on all scales, in all places, and at all times. However, limits of this generalization have already been encountered in the possibilities of extreme conditions, such as in black holes. Nature may defeat our expectations in the realm of the very complex as well as at extremes of scale. Mechanist thought is about simple linear causes and effects. It is little wonder it does not readily accommodate the complex feedback loops, non-linear processes, and multiple causation involved in self-organization.

While it seems common sense that everything must have a cause, the very notion of causal connection or causal process is merely a cipher for the faith, derived from common experience, that we can discover some "mechanism" to account for observed patterns. Where we cannot discover it, we are nevertheless compelled to imagine it exists. For the world is to us essentially a black box, to which we attribute inputs and outputs and speculate on what plausibly accounts for them. The modern idea of *natural* cause is basically efficient cause—some intrusion from outside a defined system. Because human consciousness can always conceive such an "outside," one expects a chain of such influences, but then wonders, paradoxically, what lies at the origin of the chain—the ultimate "outside." Intuitively, the notion of first cause remains animistic: the chain must stop with a responsible agent, whose initiative is uncaused *by definition*. "Common sense" thus reflects the traditional dualism of mind and matter: in the present terms, a dualism of first cause and efficient cause. This owes partly to our psychology and general philosophical tradition and partly to the mechanist philosophy in particular. Moreover, self-organization could not appear significant as a phenomenon to be explained as long as organisms were simply and naively perceived as free agents—that is, until they were first conceived as mechanisms subject to efficient cause.

In a deterministic universe, there is no room for free agents, who must somehow occupy a place outside the system (ultimately, outside the physical universe). Thus, appeals are made to a nonmaterial preserve of free will: to God as first cause to set the deterministic cosmos in motion, and to *mind* or *soul* to set the deterministic body in motion. Animals had been regarded as autonomous agents before Descartes and La Mettrie, who cast the pall of determinism over them and on the human body as well, turning autonomy into automation. Within the religious context of the time, humans could retain free will by virtue of their nonmaterial souls—a courtesy not extended to other creatures, and certainly not to machines. Yet, like consciousness, free will itself remains unexplained within the mechanist framework. Only in the last half-century have consciousness and will begun to be (re)considered within the framework of embodied agency.

If it still seems mysterious to us that we are examples of matter that can see and feel, from a perspective of its own, this is in large part because of the simplistic and idealized view of matter we retain as passive and inert, consisting of closed isolated systems, corresponding to simple mathematical models. Since the time of the Greeks, at least, we have viewed matter with condescension, as categorically inferior to our mental and spiritual status as free agents. The other side of that coin is the inflated view we have retained of our own spiritual status, as above or transcending physical matter. The scientific view of the world must now expand to include a concept of agency appropriate to what is still considered inanimate matter. Self-organization is key to further advances even in cosmology, and to the resolution of conundrums such as the apparent highly improbable fine-tuning of the universe to life. But to approach self-organization lucidly, science must confront its persistent dualism, manifest in the separation of observer and observed, theoretician and theory.

Why does it seem that nature can be grasped through relatively simple laws and mathematical formulae? The old answer was that nature was comprehensible because it was the creation of a rational mind, like ours, which was kind enough not to hide its workings from our view. A contemporary answer is that the world is comprehensible because it is inherently a "mathematical structure." Such a view is no less dualistic. A more fruitful answer, I believe, is that the world is comprehensible because it has powers of self-organization mirrored by the self-organization of the brain that contemplates it. That is, the cosmos is comprehensible because we are made in *its* image, not in the image of a God separate from it. We must come to acknowledge and understand the differences between an artifact created by external agency and a self-creating, self-maintaining, self-defining system, such as we ourselves are.

The human mind is adept at a kind of speculation driven by the need for tidy principles, categories, and relations. This need for self-contained systems of thought is a core ingredient both of theology and of scientific theory. Both presuppose the separation of mind and body. The scientist, as creator of theory, shares with God, the creator of matter, a spiritual or mental identity different in essence from the material objects of study, as well as a point of view outside nature.

The mind-body schism was reflected in the divergent paths of mechanism and vitalism in life science. Vitalists held that something nonphysical was required to explain life. Modern thinkers would call that something *organization*. But, one must ask, organization according to whose scheme? A mind outside the system is still required, whether to create or to understand its order. The mind-body schism inheres in the very concept of mechanism, which implicitly requires a designer and a mechanic, just as a computer requires a programmer and an engineer, as well as a user.

Reconciling the immanent reality of nature to the biblical account of creation occupied generations of theologians (some of whom were scientists) in the wake of the Scientific Revolution. The struggle, however, was not so much over the autonomy of nature as over human autonomy. It was a struggle between determinism and free will; or, between God as master of human minions and men as aspiring masters themselves, who would have other minions under them—whether slaves or machines. The philosophy of mechanism paid lip service to divine power while underwriting human power. Asserting that organisms are machines was a double-edged sword, however. Humans exempted themselves from unpleasant implications on the grounds that their bodies served only to house, but did not define, their spiritual essence. Otherwise, the implication was that all organisms could be understood *as though* they were human artifacts, constructed of the same inert matter that serves as raw material for industry. The possibility that matter is *not* fundamentally inert was overshadowed by the practical usefulness of the mechanist stance, which empowered men to do what they would with nature. The possibility that organisms had somehow constructed *themselves* was upstaged by the promise that bio-engineers would one day do it.

The Nature of Agency and Agency in Nature

One might expect a conscious social primate to view the world in terms of agency. Indeed, we are not surprised by animism and the broad projection of intentionality into nature for most of human existence. In view of the naturalness of the notion, it is all the more remarkable that modern science shuns it in "naturalistic" explanation. There are two historical reasons for this rejection. The first follows the ancient Greek mandate to expunge the supernatural from rational thought. The second is the concentration, in the Semitic religions, of powers of agency in a monotheistic creator god external to the world.

A nature with immanent reality must have the autonomy associated with organisms. While organisms are seats of intentionality, and may be said to have their own purposes, the standard for "intentionality" remains human consciousness, purposing, and creativity. Of course, natural systems of all sorts often lend themselves to analysis in intentional terms.[5] We try to understand the functions and parts of even inanimate things in terms of human purposes, design concepts, and engineering principles; science is effectively a process of reverse-engineering nature. However, we run into trouble in trying to think of the organism as *self*-engineering, when by engineering we have in mind only the top-down creative processes that we think of as characterizing human intentionality.

While it may be our habit to conceive of agency in human terms, concepts of agency and intentionality could reasonably be broadened to include any convergence of means toward end, including the adaptations of organisms to their environments. The brain, for example, is plausibly an agent. The internal connections within it, while physical, can also be described as intentional, since they converge toward ends and often involve the sort of "directedness" associated with intentionality. If the brain can be viewed in both causal and intentional terms, why can't other complex physical systems be viewed the same way? The brain may be a bridge for understanding general self-organization in the universe, because we associate with it both our own first-person point of view and third-person description; we accept it as a causal *or* an intentional system.

To assume chance to be the only operative factor in evolution makes life seem an unlikely affair. Even today, this leads some to believe that life could only arise by "design." Yet both chance and design overlook the processes by which life shaped the conditions for its own development on this planet. Consideration of similar processes at work in the cosmos at large, if such exist, could reduce the apparent odds against the brute chance occurrence of an orderly, life-bearing universe. Some processes of cosmic self-organization are currently recognized—in galaxy formation, for example. Broader questions include what role self-organization plays in leading to the kind of universe we live in, and whether it leads inevitably to life and intelligence.

Reproduction is such an obvious feature of life that natural selection of mutations over generations is considered the engine of evolution. Yet reproduction is only one of several defining characteristics of life, which also include self-definition, self-organization, and self-maintenance—broadly speaking, self-creation, or autopoiesis.[6] It may be unnecessarily limiting to think only of some version of natural selection as the basis of cosmic self-organization.

The challenge is not to show how "blind chance" can produce emergent phenomena that look designed. It is rather to show how natural phenomena that look like blind chance can be viewed as the result of natural agency. Toward that end, we labor against a longstanding psychological investment in our identity as active agents, in contrast to allegedly passive matter. If we seek a mature understanding of our place in nature, however, we must cease to demean nature's power in order to assert our own. Accordingly, one goal of a revised science would be to demonstrate that such an apparently unlikely cosmos would be nearly inevitable in the light of deepened understanding of processes of self-organization. In the meantime, it is feasible to naturalize concepts such as "mind," "intentionality," and "agency" so that they may apply to processes still regarded as mechanistic and passively inert. At present, neither religion nor science recognizes natural agency, and the concept of *self-design* has yet to be explored in depth.

The assumption that nature is comprehensible because it was created by a rational being endorsed the right to project human thought processes back upon it. One could assume, alternatively, that nature creates *itself*—though not in the top-down manner of human creativity. Nature would then be comprehensible because we are part of it, rather than because we are different and separate from it. If matter is intrinsically passive and disordered, an external intelligence is required to impose order upon it. But if matter creates its own order, then the order of human reason must exemplify it.

Design by an external agent has been an important working concept in the history of science, sometimes subtly. One can pay lip service to self-organization, for example, even while holding that it remains "guided" by transcendent laws. Yet a self-designing cosmos could explain the appearance of intelligent design without assuming the separation of body and mind into creation and creator, into machine and engineer, or into physical systems and transcendent mathematical laws. Nevertheless, science has tended to side with religion against a self-designing nature, in favor of a view of nature that imposes order upon it from without. The external agent is no longer God, but the theorist. Even today, a final theory is attractive for reasons similar to the appeal of the absolute truths of religion. Both hold that the ultimate cause of physical things does not lie in the world itself but in some transcendent principle.[7] Perhaps this reflects a fundamental masculine preference for transcendence over immanence, and it may well be that science will not free itself from such metaphysical assumptions until it recognizes this tacit gender bias.

The Scientific Revolution involved a deliberate shift from intentional agency to efficient cause. Nature was to be an implicitly passive object to the creative human or divine subject. A modern echo of this duality is found in the dualism of laws and initial conditions. Matter passively "obeys" mathematical laws, but the initial or boundary conditions must be written into the equations by hand, whether human or divine. For the early physicists, God played the obliging role of setting initial conditions as well as decreeing laws. More secular theorists later felt free to specify these inputs themselves, with the implication they could also freely create alternative "worlds" on paper. The notion of law is still widely considered transcendent rather than inherent in the stuff of the world.

This shift away from agency was the basis for the mechanist vision and the banishment of everything subjective from scientific discourse. Accordingly, the concept of agency went underground, so to speak, preserving mind-body dualism. It allowed mechanical interactions within simple systems, on the one hand, and required an external first cause, on the other. That cause could be God, the creator of the universe; alternatively, it could be the theorist, who sets initial conditions of a mathematical system, the experimentalist who defines the inputs of an apparatus, or the programmer who feeds the input to a computer program. In all cases, the intentionality involved remains separate from the system itself. (It is even separate from the brain—leaving phenomenal experience, selfhood, and free will without a scientific explanation.) More significantly for the history of physical science, it also channeled analysis away from concepts of agency and teleology essential for understanding the self-organization of complex systems.

Aristotle had likened genesis in nature to artisanal production: a program of deliberate steps. This kind of anthropomorphic projection underlay the creationism of the early scientists, and continues to underpin the notion of physical law as algorithm. Mechanism took the denaturing of nature a step further. The world was now a blank raw material for human consumption, parallel to Locke's tabula rasa.

While the various kinds of cause could coincide for Aristotle, they were separated in the new mechanist view. To be sure, God had his purposes in mind for the functioning of each natural thing and its parts, as well as for the Creation as a whole. But either it was presumptuous or futile to try to fathom these (Descartes and Vico) or else the proper role for such inquiry was to magnify faith (Boyle and Newton).[8] Either way, scientists were left with an inert and mechanical cosmos, while teleology lay outside the purview of physical science until the development of cybernetics in the wake of the Second World War.

In early twentieth century, the new quantum theory shattered the Newtonian worldview, while maintaining an ambiguous relationship to determinism. (The wave function is deterministic, for example, while its "collapse" is unpredictable.) On the one hand, even the actions of the human organism are supposed to be determined finally by underlying physical causes. On the other hand, at the quantum level, in the last analysis physical causes remain undetermined. Einstein expressed his dissatisfaction with this inconsistency in what was perhaps more than a trope: "I find the idea quite intolerable that an electron exposed to radiation should choose *of its own free will*, not only its moment to jump off, but also its direction."[9] This sentiment well expresses the deterministic view of nature as inert and passive, and the requirement that it be subject to prediction and control. At stake is the possibility that, if nature is unpredictable, then effectively it is an independent agent, to which we logically should accord some counterpart of what we call free will.

Ultimately, one must ask, what is the difference between the indeterminism of nature at the micro level and the indeterminism of free agents? What, after all, is the difference between the acausality we attribute to random natural processes and the acausality we attribute to first causes, to human or divine whim? Though for different reasons, both elude understanding and prediction. Whether we want to control electrons or people, we can only do so through some form of determinism: that is, by identifying causes that guarantee effects. But if the only truly deterministic systems are deductive systems (namely, our own creations), wouldn't it be more honest to admit that nature is indistinguishable from a creative agent in its own right, not so different from ourselves? When viewed third-personally, free will is an emergent phenomenon, arising from apparently deterministic processes. Yet it is also the result of a particular way of categorizing things. We choose to regard as free agents the complicated tangle of tropisms we call human beings. This involves not only an ontological claim about the human organism, but also a special relationship to it. Nothing in principle prevents us from bringing a similar relationship to other parts of natural reality.

Information in Living Systems

As an aspect of human affairs, information is inherently semantic, referring to conscious agents and meanings. The plausible context of a causal role for information is therefore *agent* causation, in which information is actively created, communicated, and used. It is not efficient causation, in which information plays a purely physical role. While the physical and chemical processes involved in metabolism, for example, can be trivially regarded as an analog form of "information processing," this adds little to the conventional understanding of these. The fact that information is transmitted through physical processes (as in nerves or telephone wires) does not render it a matter of efficient cause only.

Genetic processes actively use digitization to transmit hereditary traits relatively without error. Similarly, nervous systems, with digital aspects, actively transmit information within the organism about the state of the organism and its environment. This requires storage separated from processing and modifiable according to circumstance. The active and explicit use of information, with dedicated storage and the capacity to update or reprogram, seems characteristic of organisms and has even been used to define life and to characterize its emergence from inert matter.[10] However, the distinction between "active" and "passive" information processing is somewhat spurious if *all* information processing is necessarily active, involving agency. The emergence of life is then not a transition between kinds of information processing but between systems that create and use information for their own purposes and systems that do not. It is misleading to attribute a special causal power to "biological information." The organism simply governs its own internal processes in diverse ways, making its own use of "information" to do so.

The situation is complicated by the fact that the path of information in organisms is not a one-way street. It is not a set of instructions from top down, as human engineering appears to be, but operates multi-directionally and circularly. The organism is not only a molar agent in the world, but consists of self-configured internally communicating parts, which may or may not be considered agents themselves. Potentially, this circumstance vastly expands the concept of agency, while pointing to the risk of projecting human agency into the organism's situation. The best we can do is to try to put ourselves in its shoes, to reason about *its* reasons.

Arguments from Design: Complexity and Teleology

Under the sway of the mechanist philosophy, science has turned a blind eye to the interconnectedness of nature that might account for teleology and apparent design. On the other hand, religious arguments for intelligent design are usually offered on the grounds that it best accounts for the complexity and apparent "teleological properties" of nature. In order to imply that organisms, like machines, were designed *by* someone, it serves the proponents of intelligent design to assume that organisms are complex machines.[11] Here we argue the opposite: such properties and the complexity of nature (whether organic or inorganic) are precisely what *cannot* be accounted for by design by any external agent. Complexity in general does not point to artifice, which is characterized rather by simplicity, ideality, and isolation of parts. Simplicity, not complexity, is the mark of top-down agency and rational design. Moreover, one must not fail to recognize the very different kind of teleology that characterizes self-organizing processes. At one extreme of conventional analysis is the impersonal agency involved in efficient cause; at the other, the agency of persons. With little in between, there remains an explanatory gap. This is the same gap that has always separated mind and matter, rendering naturalistic explanations of mental phenomena persistently challenging.

Natural systems might inspire a machine's design, but someone designs and builds it. It has a finitely delineable structure, the precisely "correct" set of well-defined parts intended by its designers. It can usually be dismantled into this same set of parts by reversing the process of construction. Whether a design is sound in principle is analogous to whether a theorem can be proven in mathematics. This is different from testing the design in practice, which depends on real materials and their behavior in real situations. Similarly, a scientific experiment is supposed to test the design of nature, as proposed in theory; yet the results depend also on the concrete realization in the experimental setup.

The mechanistic view of nature applies the machine metaphor not only to isolated systems but, more broadly, to nature at large—which one then assumes can be analyzed into its "true" parts, in the reversible way that an engine can be assembled and disassembled. This assumption reflects and supports the religious view of the cosmos as an intentional creation. In a book published in 1802, theologian William Paley proposed an argument for the existence of God that has since been known as the "watchmaker argument." (Plato had given an early version of this argument in the *Timaeus*.) If you were out on a stroll in the woods and you came upon a pocket watch lying on the ground, he reasoned, you would think of it very differently than of a stone lying there beside it. The watch stands out as having a rational design, and must therefore have had an intelligent maker and a date of manufacture. He then goes on to assert that the works of nature similarly bear the marks of design that indicate an intelligent creator: the stone itself should be viewed as an artifact! That is hardly a valid conclusion, of course; it unwarrantedly projects human intentionality back onto nature through the mechanist stance. In a wily inversion of the watchmaker argument, one critic of the mechanist view of life has likened it to taking a hammer to a watch.[12] A hammer can certainly reduce a watch to a variety of pieces, but these will not likely be the original "correct" components, and the exercise will shed little light on the structure and functioning of the parts, the nature of the whole, or the process of fabrication. In the case of the organism, we can only guess about these things, since (excluding God) no one made it in the first place.

Most scientists and philosophers readily admit that nature abounds with "dazzling intricacies" that appear rationally designed.[13] Even staunch opponents of creationism marvel at the intelligence displayed in nature, just as many marvel at the unreasonable effectiveness of mathematics. Yet, as we have seen, math works so well in science partly because it generalizes features of nature to begin with. Similarly, the effectiveness of the "design stance" derives from the fact that design principles (and even rationality itself) have largely been abstracted from nature to begin with. Nature produced reason, not the other way around. Thus, one should not be surprised at features of nature suggesting rational design.

Design, like abstraction, means idealized simplification. Nature's dazzling intricacy is evidence *against* such simplification. Thus, while we marvel at natural intricacy, it is precisely *not* a hallmark of design—if by that we mean the organization imposed externally by a conscious designer. Design arguments typically reason from the premise that a natural thing bears certain characteristics in common with a made object. From this, one is supposed to conclude that it too was made. However, the characteristics in question are themselves artifacts of human analysis, so that one product of definition is being compared to another, rather than to a natural thing. Paley's watch, like any machine, bears the marks of contrivance by virtue of its pristine character as the embodiment of a deductive system. For this very reason, it resembles nothing natural; for nature is not simple and tidy but complex and ambiguous. Furthermore, the very concept of an external designer invites one to ask: who designed the designer?

The complexity of artifacts does not resemble the complexity of natural things. Artifacts are not autonomous in the way that natural things are, whether living or not. Such teleology as artifacts have is conferred by their designers. If we marvel at the intricacy of a watch, it is because we know the difficulty of making such a thing. If we see the "same" intricacy in nature, it is because we are incapable of seeing further into natural complexity than our mechanistic thought permits—the kind of thought that could produce a watch but not a tree. The intricacy of nature may infinitely exceed that of human design. It is as qualitatively different as the found is from the made.

We have *learned* to discern both the complexity and the simplicity of nature through eyes that have been trained to see parts and relationships between them. However, these may be artifacts of our habits of thought, often reflecting the design principles of made objects. The true (and perhaps indefinite) complexity of nature may be hidden by our very ways of looking. If our level of comprehension mirrors the sophistication of our artifacts, then we see only the depth of complexity we have been able to create. In any case, complexity and interconnectedness do not support creationism, but point rather to powers inhering within nature itself.

The dualism of inert matter and governing laws, which grew out of the ancient distinctions between substance and form and between object and subject, cannot account for the evolving complexity of the physical universe. This first became apparent in the life sciences and in conflicts between the biblical account of Creation and archaeological, fossil, and geological records. The problem was (and remains) to explain how the complexity of life could arise from the passive inertness of inorganic matter. In particular: how a mechanical system could organize itself in such a way as come to life and consciousness. Put this way, we see that the problem is rigged from the start. For "mechanical system" and "inert matter" are *assumed* to lack the required properties in the first place. With such assumptions it is equally difficult to imagine, for instance, how the suite of "symmetry breakings" that are held to account for the particle zoo could take place "spontaneously" or "unguided" by preexisting laws. Yet how can we explain the prior existence of laws any more easily than the prior existence of a potential for self-organization? It may seem facile to credit the evolution of matter to an inherent potential within it. Yet one should not interpret a deficiency in thought as a deficiency in nature.

Mechanism

The philosophy of mechanism is based on a conception of matter as passive. In contrast to earlier mechanists, however, eighteenth-century thinkers incidentally transferred back to nature powers that had been ascribed to God. Just as Laplace had no need of the God hypothesis to account mathematically for the behavior of nonliving matter, so La Mettrie concluded that God is not needed to account for the spontaneous activity of living matter. While this was not a question of resigning nature's order to chance, but to laws inhering in nature rather than attesting to divine will,[14] the metaphysical standing of such laws remained unclear.

Descartes "dangerous idea" had been that the organism, and indeed the whole of nature, is a machine.[15] The corollary he drew is that what is *not* a machine is the human soul animating the body, along with the whole of a spiritual realm animating the physical world. Leucippus and Democritus had earlier perceived the cosmos as lifeless machinery in which a strict necessity rules over constituent atoms.[16] They had attempted to purge cosmology of intentionality, leaving only impersonal, objective events. This was the first deliberate separation of the mental and the physical in Western traditions, and the first reduction of the diverse world of appearances to an underlying mechanical order. It paralleled the kind of deductive thought that found contemporaneous expression in Euclid's *Elements*. Ironically, the atomists "eliminated" the mental from their cosmology by translating the physical into their own mentalist terms.

The extent to which the Renaissance had rediscovered atomism as the mainspring of the world may be judged by this passage from *Tom Jones*: "The world may indeed be considered a vast Machine, in which the great Wheels are originally set in Motion by those which are very minute, and almost imperceptible to any but the strongest Eyes."[17] Similarly, Leibniz wrote of living beings as "machines of nature" whose key feature is that they are mechanical down to their smallest parts.[18] Yet, as I have argued, machines are categorically *simple*, so that the notion of an infinitely complex machine is effectively an oxymoron.

The geared clock served as the precursor of the machine as motor, the paradigm of the Industrial Revolution. The essence of the machine as a deductive system could be exploited in calculating "engines," such as Leibniz toyed with. Yet a machine—even the machinery of life—needs a mechanic. For most of European intellectual history, an active principle of some sort was assumed to animate the physical body, just as the human programmer animates the computer.

Though an infinitely skilled "clock maker" may have appealed to the Enlightenment imagination, any mechanism remains a finite product of definition, if it is to have any sense at all. Only since mid-twentieth century have machines been conceived that could regulate and repair themselves, perhaps even with the potential to self-construct and reproduce. Only relatively recently has the divide between organic and inorganic ceased to have a clear ontological meaning, with the result that *organism* can now be taken seriously as an alternative to both mechanist and vitalist or spiritual notions. For the characteristic of organism is complex interconnectivity. In that vision, the world is a network of interactions, integrated and dynamic, in which parts both affect and are affected by all other parts, and can only be adequately defined in terms of their *self*-integration within the whole. In contrast to the rest of science, it is ironic that biology, at least in the guise of commercial gene technology, continues to follow a mechanist model.[19]

Organism

Nature as a whole was once viewed as having the kind of autonomy we associate with livings things. Like an organism, it was thought to have powers of agency, not merely the passive inertness to which mechanistic science has reduced matter. The cosmos as a whole was considered self-existent. Despite the religious view that nature has only a derived reality, in late medieval Europe the world was still conceived to be alive at many levels. Even metallurgy and geology were conceived in organic terms, with harvestable veins of metals growing in the earth from seeds implanted from the stars.[20] Nicolas of Autrecourt had expressed the sentiment, which resonates with modern ecological understanding, that "the beings of the universe are connected among themselves, so that . . . nothing exists unless it is good for the whole multitude of things that exist."[21] From the Greeks on, however, there had begun a relentless slide into a more fragmented world, with neither its own agency nor immanent reality, but reducible ultimately to human systems of thought. Based on the separation of mind and matter, the machine gradually displaced the organism in the public and scientific imagination.[22] Yet organism has a venerable history as the context in which a mechanistic science could arise.

Following Aristotle, matter had been the passive recipient of form in medieval thinking. With seventeenth-century science, matter continued to be passive, while form was divinely imposed. Matter was no longer Aristotle's vague substance but something that could be definitely known and manipulated, having specifiable properties. For the ancients, material properties had been humanly asserted attributions. The new scientists came to view them as divinely fixed and independent of human thought.

Aristotle's notion of science, as the study of the individual specific "natures" of things, strongly influenced the medieval concept of nature at large. Powers within natural things themselves constituted the source of change. In contrast, artificial things possessed no such inner nature but were products of external agency. In rejecting Aristotle, the new science retained only the "efficient" causes that operated *between* things held implicitly to be artificial and passive. The type of cause originating in specific natures—that is, in the immanent reality of natural things—was transferred wholesale to the divine will. Moreover, Aristotle had held that the aggregate of natures constituting a complex thing is not simply the sum of its parts, but constitutes a distinct nature characterizing the unified whole. While this seems obvious in the case of life, the reductionist philosophy rejected this teaching, which is only recently being recovered in the notion of emergent properties. In general, nature was merely the sum of its parts, without self-organizing powers of its own.

While Aristotelian "natures" may lack explanatory power in modern terms, they may yet prove useful in thought about the complex interrelationship of things involved in self-organization. Galen had early noted that the difference between health and disease lies in the body's homeostatic stability, its ability to resist disintegration. A healthy body resists relatively extreme disturbances, whereas an unhealthy body may respond with dramatic effects to relatively small disturbances.[23] In special circumstances, robustness or stability may indicate a simple closed system, passively isolated from disturbing influences. More widely, however, it may signify a complex homeostatic system able to *actively* compensate disturbances. In the one case, stability is incidental and artificial; in the other, it is systemic and "teleological."

So far, human artifacts in general, and artificial intelligence in particular, do not have that kind of robustness. They are not self-generating or self-maintaining, but are accessories to human purpose. They are signs in a network of human meanings and cannot easily be separated from the matrix of culture. The first reading of nature in Christian culture was accordingly human-centric. While this stands in contrast to present understanding, the idea that nature could have its own teleology—indeed, hold its own values and point of view—remains foreign to the entire heritage of modern thought.[24]

It can be argued that there is no such thing as *an* organism. Individuals are never truly independent of a larger context. Cell biology may study the part, arbitrarily ignoring the integration of parts into larger wholes,[25] but parts without context are effectively artifacts. No living form can properly be defined as a self-contained unit nor only in terms of its individual reproductive capacity or history. It is only under conditions of artificial isolation from the rest of the organism or population that biological phenomena appear to involve simple causation.[26] The very notion of *stimulus-response* in laboratory situations artificially treats the organism as passive, separating it from its own participation in managing "stimuli."[27]

Every creature is not only the product of its ancestors but also of the whole web of life. Species coevolve with other species; the biosphere evolved as a whole system. Something similar may be true in the inorganic world as well. Quarks, for example, can only exist within a larger whole—the proton. Electrons "cooperate" with each other in the exclusion principle and in superconductivity.[28]

Via the historical bridge of mechanism, our era is regaining an interest in nature's inherent powers of self-organization. The model for nature was first organic, then mechanistic, then computational. The next best metaphor to model the complexity of nature may again be organic: the brain. Even mathematics may benefit from a more organic model—for example, axiomatic systems that evolve dynamically.[29] In any case, physics and biology represent complementary and mutually informing windows on the world. Genetic engineering notwithstanding, the lesson to draw from biology may be to try to understand the world on its own terms rather than to manipulate it on ours.[30]

Notes

1. Smolin, Lee. 1997. *The Life of the Cosmos*. Oxford: Oxford University Press, p. 190.
2. Kingsley, Charles. 1890. *Water Babies*, London: Macmillan.
3. Laughlin, Robert. 2005. *A Different Universe: Reinventing Physics from the Bottom Down*. New York: Basic Books, p. 31.
4. Debus, Allen G. 1978. *Man and Nature in the Renaissance*. Cambridge, UK: Cambridge University Press, p. 94.
5. Hence, Dennett's *intentional stance*. See Dennett, Daniel. 1987. *The Intentional Stance*. Cambridge, MA: MIT Press.
6. A term coined by Maturana and Varela. See Varela, Francisco J. 1992. "Autopoiesis and a Biology of Intentionality." In *Proceedings of the Workshop on Autopoiesis and Perception* edited by Barry McMullin. Dublin: Dublin City University.
7. Smolin, *The Life of the Cosmos*, p. 199.
8. Osler, Margaret. July 1996. "From Immanent Natures to Nature as Artifice: The Reinterpretation of Final Causes in Seventeenth Century Natural Philosophy." *Monist* 79:388–407.
9. Albert Einstein, in a letter to the Borns, April 29, 1924, quoted in Earman, John. 1986. *A Primer on Determinism*. Dordrecht, NL: D. Reidel, p. 199.
10. Walker, Sara Imari, and Paul C. W. Davies. 2012. "The Algorithmic Origins of Life," p. 6. http://arxiv.org/pdf/1207.4803v1.pdf.
11. Brigandt, Ingo. March 2011. "Explanation in Biology: Reduction, Pluralism, and Explanatory Aims." *Science and Education* 22:69–91.
12. Rosen, Robert. 1991. *Life Itself*. New York: Columbia University Press, p. 22.
13. Ratzsch, Del. June 2005. "Teleological Arguments for God's Existence," section 2.2. http://plato.stanford.edu/entries/teleological-arguments/.
14. Roger, Jacques. 1986. "The Mechanistic conception of Life." In *God and Nature*, edited by D. C. Lindberg and R. L. Numbers. Berkeley: University of California Press, p. 288.
15. The clockwork universe was even first thought to be *designed* to run down on a timetable leading to the apocalypse. See Webster, Charles. 1982. *From Paracelsus to Newton: Magic and the Making of Modern Science*. Cambridge, UK: Cambridge University Press, p. 21.
16. Lindberg, David C. 1992. *The Beginnings of Western Science: The European Scientific Tradition in Philosophical, Religious, and Institutional Context, 600 BC to AD 1450*. Chicago: University of Chicago Press, p. 31.
17. Fielding, Henry. *Tom Jones*, Book V, Chapter IV.
18. Wolfe, Charles T. 2009. "Why Was There No Controversy over Life in the Scientific Revolution?" p. 11. PhilSci Archive. http://philsci-archive.pitt.edu/5409/1/cw-_Life_in_the_Scientific_Revolution.pdf.

19. Suzuki, David, and Holly Dressel. 1999. *From Naked Ape to Superspecies.* Toronto: Stoddart, p. 105.
20. Debus, *Man and Nature in the Renaissance*, p. 34. Metals were mistakenly considered a renewable resource, like wood.
21. Circa fourteenth century. Quoted in Franklin, James. 2001. *The Science of Conjecture: Evidence and Probability before Pascal.* Baltimore: Johns Hopkins University Press, p. 213.
22. Even the ancient Greeks had dreamed of robots—perhaps because they were a slave society. On a more fundamental level, they had discovered the power of deductive systems, which is the basis of mechanism.
23. Barrow, John D. 1991. *Theories of Everything: The Quest for Ultimate Explanation.* New York: Fawcett/Balantine, p. 168.
24. Hence, the problem of how to explain the emergence of complexity from "simple" systems parallels the problem of how to explain the emergence of mind from matter and the emergence of irreversibility from reversible processes.
25. Capra, Fritjof. 1996. *The Web of Life.* New York: Anchor/Doubleday, p. 24.
26. Goldstein, Kurt. 1963. *The Organism.* Boston: Beacon, p. 419.
27. Varela, "Autopoiesis and a Biology of Intentionality," p. 9.
28. Barbour, Ian G. 1990. *Religion and Science: Historical and Contemporary Issues.* New York: Harper, pp. 174–75.
29. Chaitin, Gregory. May 2007. "Algorithmic Information Theory: Some Recollections." "Challenges for the Future" section. www.cs.auckland.ac.nz/~chaitin/60.pdf.
30. Woese, Carl R. June 2004. "A New Biology for a New Century." *Microbiology and Molecular Biology Reviews* 68:173–86.

18

Theories of Something

> *"We cannot expect everything of a Theory of Everything."*
> —John D. Barrow[1]

> *"It is only myth that attempts to say how the universe came to be, either four thousand or twenty billion years ago."*
> —Hannes Alfvén[2]

Introduction

Generations of scientists have dreamed of finally extracting the last of nature's secrets, sometimes deluding themselves that the horizon of knowledge has already been reached, and that all that remains are clerical tasks of filling in the details of a definitive understanding. This sense of "closing in" on nature hinges on the assumption that natural systems can be corralled in a set of definitions, as though they were mechanisms whose blueprints could be reverse engineered. What makes nature *real*, however, is that it contains, engineers, and defines *us*, and not the other way around.

A theory of everything "would be a set of simple, elegant equations that give you the whole universe, and which would fit on a T-shirt."[3] It would be "final" in the sense that it would "bring to an end a certain sort of science, the ancient search for those principles that cannot be explained in terms of deeper principles."[4] It would consist of definitive laws of physics, with a minimal set of initial or boundary conditions.

Initial or boundary conditions of a system are not implied within equations themselves but must somehow be specified. They may be set by the history or the environment of the system, by chance, or by the theorist. In the case of the universe as a whole, of course, there is no environment and no history to set such conditions. This has led many cosmologists to speculate that the universe we perceive might be but one tiny "bubble" in a much vaster context, a plurality of universes. If such a "multiverse" were large enough, perhaps infinite or eternal, the problem of initial conditions would disappear, simply because enough trial runs of random "universes" would inevitably produce one like ours. Many scientists are not satisfied to let the beginning go unexplained, and some question the extravagance of bringing in a multiverse to contend with improbable initial conditions, simply deferring the issue of the origin to a larger context. Moreover, chance can only be eliminated if initial conditions are either *derived* from theory or are somehow rendered *irrelevant*.

In effect, this book has argued that an initial state could be theoretically derived only in a deductive system. On the other hand, the multiverse idea turns chance on its head but at a high metaphysical price. However, initial conditions would be irrelevant in a universe that is somehow *insensitive* to them. That is, it would follow a similar path of evolution regardless of the initial state. A nature with general powers of self-organization might well converge toward some configuration in a manner prescientifically characterized by "final cause." Its state would not simply be a domino effect of some prior state.

The notion of possible universes arose in medieval discussions of Aristotle, who taught that the world can only be as it is, according to its own inner nature and logical necessity. The Church saw in this a denial of God's omnipotence. This led scholars to also reject Aristotle's teaching that there is no void, since there had to be some vacant space in which God could exercise his power to create alternative worlds. Hence, they affirmed the idea that there must be a meta-space in which the universe could meaningfully be said to move, rotate, and exist as one possibility among infinite others. While Aristotle held motion to be relative to other bodies, Newton later found sanction here for a concept of absolute space.

The very notion of another universe, of course, is paradoxical for an empirical science, since by definition we normally mean by "universe" all that can be seen or that can causally interact with us.[5] Moreover, the notion of cosmic natural selection, operating in some meta-verse, presumes a mechanism that generates universes—perhaps with minor variations on the model of genetic mutation—and a mechanism of selection, perhaps on the model of biological evolution.[6] Arbitrary assumptions must be made about such processes and their hypothetical venue.

Neo-scholasticism

In the rush to find a final theory, contemporary physics often seems to indulge in such wild and untestable speculations as to come full circle, resembling the gothic scholasticism that science was conceived to displace in the first place. The hallmark of such scholasticism is that it is a self-contained deductive system. Indeed, model universes are literally computer programs, and natural laws are now often regarded as algorithms that govern the evolution of natural systems in the way that a computer program governs the evolution of the states of a computer. This is a flawed metaphor, unless one is willing to assume that physical reality is literally a deductive system in the first place.[7]

The deductionist program in science was first fully expressed in Newton's *Principia*, presented as geometric proofs in the style of Euclid. It is inherent in Einstein's later thought, whose confidence in mathematical formalism was inspired by his success with general relativity.[8] It is encouraged by textbooks, which teach physics in terms of logical rather than historical development. This revisionist approach makes the laws of nature seem falsely simple and inevitable.[9] It also creates the impression that nature itself can be axiomatized, in a final story that has erased its tracks and all traces of historical process, dispute, or epistemology. A deductive system is timeless, eliminating dependency on the particulars of the real world. As we have emphasized, however, the reality of nature consists in this particularity, implying unaccountable change and new discovery.

Though useful and essential as a thread in science, the deductionist program leads to imbalance if it goes too far beyond empiricism. Thus, cosmologist Lee Smolin warns of a crisis in fundamental physics in which mathematical consistency replaces experiment.[10] No doubt, the crisis stems in part from the mounting costs of experiments since research in "fundamental" physics means access to ever-higher energy levels. Mathematical consistency is a lot cheaper than experiment! The definition of fundamental physics (concerned with discovering the ultimate laws of nature) is also subtly prejudiced against the very domains of study that might help resolve the crisis: materials science, biology, complexity and chaos, self-organization, etc., which are not considered to involve a search for basic laws, but only to apply known laws, often for practical benefit. In contrast to modern theories of everything, which are highly abstract, the future of science may hinge on the more modest example of materials science, which is highly empirical, reveals processes of self-organization, emphasizes emergence, and demonstrates phenomena unpredictable from first principles, such as occur at phase changes.[11]

We have already examined the religious origins of the deductive program, which systematically confuses theory with nature itself. This program continues to manifest in the quest to derive physical laws, physical constants, and even initial conditions from fundamental theory. It lies behind the wishful thought that the universe itself is a computer, simulation, or subbranch of mathematics. A radically alternative approach could lead to a theory of origins that is robustly insensitive to initial conditions in the way that life is, and which is realistic in the sense that it would rely only on empirical evidence and would not attempt to reduce nature to some idealist scheme.

In Search of a "Complete" Theory

No one expects every detail of the actual world to fall out from theory, which the phenomena of chaos and quantum indeterminacy alone would preclude. But physicists and cosmologists do expect a "complete" theory to unify the basic entities, forces, and laws of physics in a single framework, an idea initially proposed by Kant.[12] Some believe it possible to account for natural constants and even the initial state of the visible universe, as well as meta-laws that might reign over the origin of any possible universe. The notion of "initial state," however, refers historically to finite closed reversible systems; it may be unidentifiable or meaningless in the case of the universe as a whole. The very notion of law ambiguously refers to identifiable patterns, which cannot be assumed to prevail at every cosmic epoch, or to the rules of a deductive system, which is timeless. The fundamental natural constants must be measured.[13] Only within some expanded deductive system might they be theoretically derived. The hope is to find this system, in which there are no free parameters and essentially one basic entity, short of which the goal is at least to reduce the number of entities and free parameters. But if nature is not a deductive system, it will prove resistant to such a reductionist program. There might always be new constants, parameters, and even laws to discover.

The ideal is a unique fundamental theory that *determines* the natural constants, the properties of elementary particles, and even the structure of space and time.[14] This is a Platonic or Pythagorean vision, signifying the venerable belief that theory has causal power, that laws of nature are fixed and "govern." It manifests through the ages in such idealist schemes as Aristotle's crystalline celestial spheres, Kepler's planetary orbits circumscribed within regular polyhedra, Eddington's numerology of basic constants, and Dirac's large number hypothesis. While such schemes may approximate observed patterns and relationships, the relation is inverted so that the real world is only a crude approximation of some mathematical ideal. Holding the ideal to be real, however, leads to false expectations. Hence, Kepler's model could not accommodate planets beyond Saturn (he also believed that the earth should orbit the sun in the course of exactly 360 daily revolutions). Eddington was seduced into claiming that the fine structure constant must be an exact ratio. Similarly, many cosmologists thought the cosmological constant should be exactly zero, and some have even marveled that the gravitational constant is not actually zero.[15] However, there is no a priori reason to expect nature to be so tidy. It is nature that must determine theory, not the other way around.

We have seen that Einstein's ideal of theoretical completeness is sheer deductionism. Not all physicists agreed with it even in his day. Max Planck, for example, took exception to Einstein's vision of theoretically determined constants.[16] Some cosmologists now ask whether the laws of nature might themselves evolve, with changing constants. Some philosophers of science suggest that the laws of nature are immanent in matter, rather than transcendent or externally imposed.[17] Though physicists and cosmologists generally assume that the values of fundamental parameters could logically take on different values, given the same laws, if those laws are descriptive rather than prescriptive then they are not logically separable from what they describe. If the universe is the only one we can know, does it even make sense to think that its "parameters" can take on values other than the actual ones? Is it reasonable to separate initial conditions from laws when we are talking about the unique history of a single universe?[18]

The determinism behind the program of theoretical completeness is the other side of the freedom reserved to the theorist. The subject is free at the expense of the object. However, the underlying dualism of subject and object reaches its natural limit when considering the cosmos as a whole, which must include the observer—whether in the role of experimentalist or theorist. To be complete, then, a theory must include the relationship to the subject, rather than focusing only on the relationships among "objective" elements. A first-order view of completeness must give way to a vision that includes the subject as well as aspects of the universe outside defined systems. This can only be accommodated within a second-order science, capable of reflecting upon its own motives, biases, and suppositions, and upon nature's freedom of action along with its own.

The Art of Twiddling Knobs

We have already noted that inscrutable divine freedom of choice in how to detail the world is equivalent to the indeterminism of nature. While natural laws may be simple and precise, the detailing of the world is complex and messy. The whim of divine will, exterior to nature and beyond human reason, is therefore equivalent to the mystery of nature's immanent reality. On the other hand, perhaps the notion of divine intervention inspired modern theorists to believe they could themselves arbitrarily fix parameters and even laws. The theorist's figure of speech—"twiddling a knob"[19]—treats initial conditions, natural constants, and even the laws themselves as variables that can be adjusted at whim. Theory is then taken to be a nomological machine whose controls can be adjusted to make a different universe.[20]

It may well be that in any life-supporting universe the relative values of the fundamental forces, of the basic particle masses and charges, and of various physical constants must be at least marginally what they actually are in our universe. We find them surprising only because we can *imagine* them arbitrarily different. Implicitly, we put ourselves in divine shoes to imagine how a machine might be set in motion from a different start, or how a different machine would operate from the same start. However, a *coevolving* universe differs essentially from such a machine.

"Universes" can be thought to vary in fundamental features such as vacuum energy density, gauge symmetries, the values of elementary forces, charges, and masses, metric signature (whether various dimensions are considered positive or negative), and even the number of spatial dimensions.[21] A *conceptual* space of any number of parameters can define a "landscape" of possible universes, in which even established power laws can take different exponents. One can then speak of a "local area" within the landscape consisting of universes resembling our own within slight margins. However, there is no physical (or even metaphysical) basis for identifying such conceptual tools with physical reality.

Fine-Tuning and Cosmic Natural Selection

"Fine-tuning" is an epithet that names a group of problems in cosmology and theoretical physics, which arise from the apparent implausibility of certain physical facts.[22] These might involve the extremely small or large values of some fundamental parameters, or ratios between them, or the apparent dependence of the actual world on critical values of certain parameters. Such problems are thought to involve "a highly constrained and implausible adjustment of the parameters of a theory."[23] The question, however, may not be why the world seems unlikely but why we should perceive it so. Why does the universe look like a fluke? The challenge is to understand why the values of these parameters seem improbable to us, or why the theorist should have to tailor them to make a theory work.

The universe we see today appears to be in a special, highly ordered state, while the second law of thermodynamics implies a continual degrading from an earlier state of even greater order. The problem is how to account for that (presumably very improbable) original ordered state. Another way to put this is that we presently labor under the assumption that the natural state of things is *disorder*. Such a fundamental premise resembles the Aristotelian belief that things move naturally toward the center. The burden of explanation falls by default upon initial conditions in the absence of a theory that tells us why there is order in a universe that is supposed to tend naturally to disorder.

Religious thinkers have embraced the concept of fine-tuning to suggest that these special conditions could not be accidental. Confusing theory with natural reality—and theoretical adjustment of equations with divine adjustment of the world—might be excusable in religiously minded lay people. However, rather than asking what is wrong with current theory, scientists too resort to forced scenarios to disarm the compelling appearance of improbability—in this case through brute numbers of trials. Just as Darwin proposed natural selection over countless generations to meet the objection that life is too improbable to have arisen on this planet accidentally, cosmologists now propose it on the cosmic scale to explain how a life-bearing universe could be a fluke. Given enough random universes, it is argued, it is likely that at least one of them would turn out to be suitable for life—and we just happen to be living in that one. Such a strategy is ontologically extravagant, entailing (perhaps infinitely) many "universes."[24] Such reasoning is widely embraced no doubt because it seems the only rational explanation available. But selection as we understand it is an essentially passive process. It may well be that nature has more active ways of self-organizing.

Cosmic natural selection metaphorically extends a familiar concept to circumstances where it cannot easily be confirmed. The motivation is laudable. Lee Smolin's black hole selection theory, for example, proposes a "time-asymmetric physics . . . that makes a universe like ours inevitable rather than improbable."[25] However, while time asymmetry (irreversibility) is necessary, it is not sufficient. There must also be processes that create the order and render it robust. A time-asymmetric universe is simply what actually exists; it is not made so by laws—or even meta-laws.

In order to accommodate his theory, Smolin develops a concept of time as "real." Such time transcends any generation of universe. It is a container for everything, like Newton's absolute space or the four-dimensional space-time manifold. The suite of universes is supposed to be causally connected within this meta-framework of time, but actually the universes are born from singularities where there is no demonstrable causal continuity. His concept glosses over this, and is no less metaphysical than versions of the multiverse with causally disconnected *spatial* regions. I believe a more appropriate question is, how did the universe evolve its own parameters within a unique cycle? But such a question cannot even be asked within current science, with its bias toward a concept of passive matter. To answer this question requires contemplating principles of self-organization beyond the mechanism of natural selection over generations.

Rather than speculate about any version of multiverse, it would be ontologically more economical to question the standard model or the ways in which it is applied to cosmological observations. Astronomy is a highly inferential science. Long chains of assumptions are involved in interpreting data—to arrive at a value, for example, for the cosmological constant. Nominally empirical measurements are highly theory dependent. The estimations of improbability inspired by them may be spurious. Indeed, the very notion of a randomly generated universe is nonsensical if there is but *one* generation of universe—one role of the cosmic die. Supposed implications of "fine-tuning" may reflect the state of science more than the state of the universe.

The idea of randomly selected parameter values (or initial conditions) corresponds to a situation of unstable equilibrium. To observe a pencil standing upright on its point begs to ask how it got into this unlikely position. One could wonder about the likelihood of such a state happening "spontaneously," or about the coincidence of the observer being present at exactly the right moment. One is at liberty to speculate that, out of a large enough ensemble of random possible worlds, such events are inevitable. But to explain the values of fundamental parameters, would it not be more plausible to look for situations of *stable* equilibrium—like a marble in a basin—attractor states insensitive to initial conditions? Rather than an explanation designed to overcome specious improbabilities, one should seek an explanation in which the nature of the system is that all possible initial conditions tend toward the state in question. We have seen, however, that equilibrium is the exception in a world in flux, and often implies an unusual or artificial situation. When there are stable conditions in the real world, most often something keeps them that way. Except in those situations singled out by classical dynamics, it is self-regulating processes that favor stability. Moreover, the relevant parameters must not be seen in isolation. It is the total package that corresponds to the attractor state. And that sort of total package is a self-organizing system.

Should one marvel that a particular course of play appears "unnaturally" selected from all possible rounds in a game? In fact, only *one* hand ever need be dealt to arrive at a unique outcome; what is natural is what actually exists. Should one marvel at a universe supporting life, when only one sequence of events is required to lead to it? The situation bears comparison to wondering why one was born the particular person one is out of all people one could have been. Just as one has to be somebody, so the world has to be some way. A more meaningful question is: What forces shaped the person one has become? Multiverses simply abdicate that sort of explanation.

Since the universe is by definition the only one we know, it is an odd idea that a different set of parameter values could have shaped a different outcome. It suggests that the experimentally found values that describe our actual world could be arbitrarily different simply because it is mathematically possible to assign different values to variables (even to "constants") of certain equations. It is the nature of the human mind to freely imagine possibilities; it is the nature of nature, however, to be just what it is and not generic or generated by an algorithm. It should not surprise us, therefore, that its given particularity could appear to us as a chance affair against a backdrop of imagined possibilities. It is theory, however, that is out of tune with reality, not the other way around. There might well exist processes through which the universe tunes itself.[26] One must also consider the tuning of the organism to its environment, which in the case of human beings might include the very thought processes that lead to the appearance of a fine-tuning problem!

Deus Ex Machina

While the accidental suitability of the cosmos for life is deemed by scientists to be extremely improbable, this is sometimes taken by others to be evidence of divine special creation. While scientists prefer naturalistic explanations, we have seen that they too may resort to extraordinary measures to accommodate them. Some scientists seem to think that we live, if not in the best of all possible worlds, then in the most "interesting."[27] On the other hand, a less than perfect world, though specially created, has long been a thorny theological issue. Theologians and theoretical physicists alike may wonder why the actual universe is for the most part empty and hostile to life.[28] Cosmologists are no more immune to just-so creation stories than theologians.

If "something" is to come into being from "nothing," it must do so either instantaneously, without cause, or else through a gradual transition from one state to the other, whatever that might entail. The traditional paradigm for spontaneous arising is divine creation by fiat. Any other process implies continuity over "time." Thus the alternative to a definite beginning is that the big bang was in some sense a change of state in some prior reality—and, hence, not a beginning at all. Even virtual particles (fluctuations) arise *within* something (the false vacuum). Ergo, either the world is natural and eternal or else it had a beginning and is an artifact, as Genesis holds!

A crucial challenge, therefore, is to understand the meaning of time outside its conventional definitions, which depend on cyclical processes that cannot exist at "the beginning." Kant held time and space to be categories built in to our perception of the world, a view that guides intuition about how to use the concept scientifically. However, we now know that such a priori categories are not logically necessary but adaptive. Our inherited intuitions about time were formed in environments that had nothing to do with cosmic prehistory, and cannot be relied upon to apply to the early universe, much less to a multiverse.

The question of meta-laws governing the evolution of laws is like the question of whether there might be higher-order terms in equations of motion: it depends on your analysis. Once time and space are defined, then velocity can be defined; once velocity is defined, then acceleration can be defined. Even acceleration may change, inviting higher derivatives. The question of how many derivatives can exist might have deep metaphysical significance without much practical significance for the observation or prediction of motion. Laws are formulated, after the fact, in terms of actual change. This is the fundamental reality, which might be described in various ways, of which time derivatives are but a particular example. The analysis of change must reflect the subject's physical circumstance and conceptual tools as much as the reality itself.

Natural selection is not the only factor in the evolution of life, and processes involved in cosmic selection may not be the only factors in the evolution of the cosmos. While it might be outrageous to think that life could shape the whole universe to its own needs, there is nothing unreasonable about the idea that cosmic self-organization could lead, if not inevitably, to life. Yet a passive view of matter is not equipped to deal even with that possibility.

Initial conditions present a double dilemma. On the one hand, they cannot be derived from theory; on the other, if they *could* be so derived it would imply that nature is nothing more than a deductive system. However, initial conditions would pose less of a problem if the present state of the world were robustly independent of them. As in the case of our planet, initial conditions of a self-organizing system might be a factor relevant to the development of life, but they would not be the whole story.

Arguments based on fine-tuning are like examining a complex machine and noticing that changing any dimension or detail even slightly would interfere with its proper functioning. While organisms, within limits, are naturally self-maintaining and robust, machines are essentially vulnerable to alteration and deterioration. This is why the mechanist universe as originally conceived required divine maintenance. There is but one way for it to work properly and an infinite number of ways for it to fail. Thus, in the mechanist universe it seems improbable for order to have come about accidentally or to have persisted through time. But what we mean by "improbable" is simply that we have no explanation other than through traditional notions relying on external agency. We have not imagined a universe that self-maintains.

It is not God but physicists who construct the machine over which to marvel like Paley. While nature is simply what it is, what perplexes us is the misleading image we have contrived of it as a delicately tuned piece of machinery. The very fact of there being laws of physics at all involves choices we make, since patterns are deliberately selected that can be expressed by computable functions. Theorists render space n-dimensional in theory, only to brood over the coincidental fact of three dimensions. They posit factors that operate in isolation and are surprised by the complexity that results from recombining them. The state of such thought at present bears comparison to the state of evolutionary theory before Darwin, when it was supposed that life (or a life-bearing planet) was virtually unthinkable "by accident." (Now, of course, it is recognized that the planet was made habitable largely by life itself!) Instead of persisting with the logic of a tinker God, it would be more profitable to take the seeming improbability of a life-supporting universe as a sign that mechanism is simply the wrong approach. We ought to inquire how diverse factors conspire together to *reduce* rather than multiply the odds against this seeming miracle.

If the values of natural constants cannot be derived from theory, perhaps it is because they do not depend on factors we can easily isolate. Perhaps there simply are no isolable functions, just as there are no isolated systems. Similarly, fine-tuning may mean simply that single factors cannot be changed significantly in isolation ("all other things being equal") without a drastically different outcome. But "single factors" characterize machines, not organisms—and perhaps not the universe. There is no reason to suppose such precarious one-dimensional dependency in nature, where multiple factors normally operate together and not in isolation. Within limits, the living world coevolves to produce a robust stability rather than a chaotic instability. That could be true in the inorganic world as well. It is a line of thought at least worth investigating. "Accident" remains in such a world, because we cannot know isolated dependencies in the manner required by determinism. This is not because the world has a mysterious fundamental property of indeterminism, but because determinism is a myth to begin with, a false expectation. There may be a role for natural selection to play in cosmic evolution, but it cannot be the only process at work. The unlikely fit of various cosmic parameters to our existence as organisms should be taken to suggest the key role of self-organization, rather than an external designer, on the one hand, or a mechanistic lottery, on the other.

Despite the example of the earth as a self-forming system, there is certainly no reason to conclude that the universe as a whole is somehow shaped by the needs of life, much less that it is in some sense alive. But this is not to say that nonliving systems cannot involve the kinds of multiple and circular causal chains that life involves. The distinction between living and nonliving systems remains a frontier still to explore. If self-organization is not a property of life alone, then it is not unreasonable that the cosmos could be self-organizing in a way that results in life.[29]

Many scientists turn to so-called anthropic reasoning to explain away the appearance of fine-tuning. Anthropic argument does not really "explain" anything, however, in a causal or reductive sense. It simply points to a condition that must be met by *any* account of the actual universe. Alone it does not account, for example, for the range of scale or the size or age of the universe, without additional hypotheses.

While the notion of a plurality of worlds is ancient, it has been widely embraced in modern attempts to explain the observed parameters of the world naturalistically, even testably.[30] It depends, however, on dubious extrapolations of concepts in quantum physics and tenuous chains of inference in astronomy, and does not even broach the question of where the multiverse comes from. It seems to be motivated by an excessive drive to leave no explanatory gap, no loose end that could—for example—open science to attack by creationists or other skeptics. Or, perhaps, it just reflects the need of cosmologists to remain employed! In any case, it may represent the last gasp of mechanism, to explain a universe that might be better conceived as self-organizing.

Notes

1. Barrow, John D. 1991. *Theories of Everything: The Quest for Ultimate Explanation*. New York: Fawcett/Balantine, p. 164.
2. Quoted in Peratt, A. L. May 1988. "Dean of the Plasma Dissidents." *The World & I*, pp. 190–98.
3. Chaitin, Gregory. "How Real Are Real Numbers?" Manuscrito vol. 34 no. 1 *Campinas* Jan./June 2011.
4. Weinberg, Steven. 1992. *Dreams of a Final Theory*. New York: Pantheon, p. 18.
5. Leslie, John. 1989. *Universes*. New York: Routledge, p. 95. Lawrence Krauss argues that, given accelerating expansion, in a future epoch the visible universe will be limited to our own galaxy! See Krauss, Lawrence M. 2012. *A Universe from Nothing*. New York: Free Press, 2012, p. 105ff.
6. Lee Smolin has proposed such a theory. See Smolin, Lee. 1997. *The Life of the Cosmos*. Oxford: Oxford University Press.
7. For instance, Svozil, Karl. 1993. *Randomness and Decidability in Physics*. River Edge, NJ: World Scientific, p. vii: "Any physical system may be perceived as a computational process. One could even speculate that physical systems *exactly* correspond to, and indeed are, computations of a very specific kind."
8. Barrow, *Theories of Everything*, p. 244.
9. Barrow, *Theories of Everything*, p. 156.
10. Smolin, Lee. 2006. *The Trouble with Physics*. Boston, New York: Houghton Mifflin.
11. Laughlin, Robert B. 2005. *A Different Universe: Reinventing Physics from the Bottom Down*. New York: Basic Books.
12. Holton, Gerald. 1996. *Einstein, History, and Other Passions*. Boston: Addison-Wesley, p. 63.

13. Among the fundamental constants are the charge, mass, and magnetic moment of elementary particles. In addition: c (velocity of light), h (Planck's constant), G (gravitational constant), and k (Boltzmann's constant). See Johnson, Peter. 1997. *The Constants of Nature: A Realist Account.* Farnham, UK: Ashgate, p. 4.
14. By virtue of the Lowenheim-Skolem theorem, however, no theory can specify a *unique* model if the universe is actually infinite. Cf. Tipler, F. J. 2005. "The Structure of the World from Pure Numbers." *IOP Reports on Progress of Physics.* 68:909.
15. Smolin, *The Life of the Cosmos*, p. 38.
16. Barrow, *Theories of Everything*, p. 122.
17. For example, Ellis, Brian. 2002. *The Philosophy of Nature: A Guide to the New Essentialism.* Montreal: McGill/Queen's University Press.
18. Smolin, Lee. June 2, 2009. *The Unique Universe.* Physicsworld.com.
19. Paul Davies, 1995 Templeton Prize Lecture.
20. For example, Weinberg, *Dreams of a Final Theory*, p. 145: "You can think of each particle as carrying a little dial, with a pointer that points in directions marked 'electron' or 'neutrino' or 'photon' or 'W' or anywhere in between."
21. Leslie, *Universes*, p. 76.
22. These problems include the "hierarchy problem" (why the various fundamental forces are of such different orders of strength); the "strong C-P problem" (why some interactions are invariant and others not, or why there is such an imbalance between matter and antimatter); the "horizon problem" (why widely separated regions of the universe seem to share the same physical properties even though too far apart for this to be a result of past thermal equilibrium); the "flatness problem" (why the universe today is poised so close to the unstable balance between indefinite expansion and re-collapse); and the "problem of the cosmological constant" (the huge difference between theoretical and observed values of the quantum vacuum).
23. Lightman, Alan, and Roberta Brawer. 1990. *Origins: The Lives and Worlds of Modern Cosmologists.* Cambridge, MA: Harvard University Press, p. 540.
24. Probability estimates for the chance arising of a life-bearing cosmos are typically said to run to 1 part in 10^{123}. To break even in terms of likelihood would require at least this number of "other universes."
25. Smolin, Lee. 2013. *Time Reborn.* Toronto: Alfred A. Knopf Canada, p. 210.
26. See Kauffman, Stuart. 2000. *Investigations.* Oxford: Oxford University Press, for a possible example: the ratcheting effect of "constructive interference."
27. For example, Davies, 1995 Templeton Prize Lecture.
28. Cf. P. Z. Myers's reply to Paul Davies's "Taking Science on Faith," *Edge*, 2007. https://edge.org/conversation/taking-science-on-faith.
29. Many macroscopic self-organized critical systems spontaneously self-organize, are scale invariant, and are insensitive to initial conditions. Cf. Smolin, Lee. May 16, 1995. "Cosmology as a Problem in Critical Phenomena." arXiv:gr-qc/9505022v1: "It is then natural to ask whether such mechanisms, or some general mechanism of self-organization, might also play a role in elementary particle physics, to explain the fine tunings, and the existence of large hierarchies, that we now must impose."
30. Carroll, Sean. December 2005. "Is Our Universe Natural?" arXiv: hep-th/0512148v1.

19

The Next Revolution in Physics?

"We cannot give the wind orders, but we must leave the window open."
—Jiddu Krishnamurti

"We had to destroy the world in theory before we could destroy it in practice."
—Ronald D. Laing

Introduction

It is perhaps ironic that the grand issues concerning the nature of physical reality—and our own nature as conscious beings—are relegated to the care of specialists in whose hands they become highly technical and literally academic exercises, largely incomprehensible to the rest of us. This situation bears comparison to that confronting the earliest scientists who rebelled against the scholastic monopoly on medieval thought, its incestuous containment in texts and purely speculative systems, its lack of direct contact with nature, and its elite language. After all, modern science emerged from the same democratic sentiment that brought about the Reformation and several political revolutions. It is commonplace that science has displaced religion as the mythos of the age. Certainly it has proposed a new creation story to replace the biblical one. Far more than fulfilling its mythic role, however, it has steered a dubious secular path to salvation through technology.

In the spirit of the Reformation, scientists turned to nature for common reference, sharing knowledge freely and equalizing observers through standard protocols and the experimental method. They sought a fresh break from the convoluted scholasticism of an elite priesthood, whose fruitless debates had little bearing on ordinary life. Modern science has come full circle, displacing religious dogma only to substitute its own hair-splitting doctrines equally removed from ordinary life and accessible only to an initiated few. Its key concepts are beyond the untrained person's grasp. Its authentic texts are in no vernacular but in the esoteric jargon of higher mathematics. It has its own priesthood to interpret its doctrines to the laity. While the scientific spirit has uncovered genuine mysteries, these are not so much answered as buried in an avalanche of technical research and technological spin-off. In some ways, this leaves society materially richer but poorer in spirit.

I believe this happens for two reasons. First, because research itself is the goal of science, rather than grand truths (of which there may few). Science is a profession (and increasingly a business), whose primary goal is its own perpetuation, financed indirectly through technological production. The important thing is business as usual. The way to sustain employment for cadres of scientists is through ever more specialization. Science, like life, is adaptive—which is one reason it survives, and also why it must be viewed as an evolutionary and social product as much as a revelation of truth.

The second reason for the upstaging of grand questions (which may, after all, remain unanswerable) is that science progresses by sidestepping them when it cannot meet them head on. One way it does this is by reifying any impasse it encounters. Just as early mathematicians did not allow themselves to be stymied by "incommensurable ratios," which they proceeded to redefine as legitimate *new* numbers, so scientists are not deterred by the apparent irrationality of nature, by the internal inconsistencies of scientific concepts, or even by lack of evidence. Consequently, real mysteries are sometimes swept under the rug—at least temporarily—such as the dual (wave/particle) nature of energy and matter, action at a distance, and quantum non-locality. Perhaps it cannot be otherwise. Perhaps we *can't* know reality in the ultimate way that both science and religion aspire to, because knowledge is only the ephemeral and relative product of very human concerns.

Of course, many scientists today *are* interested in the grand questions and do take their outreach opportunities seriously. Many use popular writing to express their private views, in plain language, about how things fit together in a larger picture outside the restrictions of their profession. Many enjoy a wide lay readership and probably read each other's books as well. There is a market for such writing because ordinary people are interested in what is happening at the "cutting edge" of research and thought. This edge, however, often takes place in highly technical realms far removed from everyday experience and concerns.

The very nature of scientific progress is to push back the frontiers of observation beyond the senses and the frontiers of theory beyond common sense. Instrument readings and hypothetical constructs replace sight and touch and everyday objects. Mathematics replaces ordinary language, as Latin did for medieval scholars.

However, something else also happens: the grand issues tend to become eclipsed by sheer technical activity, whether in offices, seminars, or laboratories.[1] The quest for truth is molded by feasible lines of research that open up, for diverse reasons, in the course of inquiry. In part, this is as it should be and may provide the maximum benefit to society; yet pursuit of truth may degenerate into pursuit of funding. While there may be few grand truths, there are many scientists, who need employment. As well as bearing the modern torch of truth, science is a social, economic, and military activity.

Perhaps it is the failure of science to live up to the Baconian promise of understanding, wisdom, and spiritual salvation—as well as secular salvation—that has led to the resurgence of religion in spite of its histories of violence and corruption. Science has not supplanted religion as many expected, perhaps for the same reasons that gave rise to science in the first place: like the medieval Catholic Church, science has become too materialistic, too removed from the vernacular, its mythos too elite, its values too self-serving.

If we hope to attain a sustainable relationship to the natural world, we must come to understand that understanding is a different goal from prediction and control. It implies a different attitude, facing, so to speak, in another direction. This has implications not only for the social deficit of wisdom, in contrast to technical mastery, but also for the future of science itself. Prediction and control have obvious survival value, but may be too narrowly focused; the scientific vision (though universal as it seems) may even threaten survival because it is ultimately too parochial and shortsighted. While defining and channeling the human relationship to nature, this vision limits the definition of science itself and the scope of research.

To ask whether science is good for humanity is like asking whether it is good to cook your food or live in a house. While these things remove us from a "natural" existence, they are what humans do. And so is science. Mankind has defined itself apart from nature, and has redefined nature too, because it is *our* nature to do so. Yet within this context there are distinctions and gradations. Food processing may be good, but this does not imply a diet of junk food. Shelter may be necessary, but this does not imply mansions and ghettos. A range of choices confront the hypothetically free will.

A Shift in Attitude

It has long been a scientific goal to explain how life could arise from the passive inertness of matter. Put this way, however, the problem is rigged from the start, for "inert matter" is *assumed* to lack the qualities necessary to account for the evolution of complexity—whether of life or in the cosmos at large.

Along with the faults that lie in the stars, we must examine those that lie in our thinking. Despite the "revolutions" of twentieth-century science, our third-person orientation continues automatically to dismiss nature as a mere "it" to be corralled and manipulated. While science and religion have traditionally agreed in treating the physical world impersonally, at this stage of the game it could be more rewarding, both in social consequences and for science itself, to relate to nature not only as an object but also *as though it were a subject like ourselves.*

Our relation to the quantum level in particular bears comparison with our relationship to living things. Both are only incidentally objects of observation. Both are inherently ambiguous. What we do know about them comes largely of treating them as products of our own definitions. Let me be clear, however: I do not propose a return to animism or nature worship, but rather a shift in *attitude*. It is not a question of what the world is or is not. The point, precisely, is *not* the ontology of nature at all but *our approach* to it and to its study. It is a question of shifting the focus toward the subject instead of the object. This is a hard lesson for a mentality that is genetically and historically conditioned to look outward at an external world rather than at the perspective from which it looks.

If this shift is to take place, the scientific view of the world must come to include a broadened concept of agency, even concerning what is presently considered inert inorganic matter. Physics is now positioned to learn much from biology. The concept of autopoeisis may be key to further advances in many sciences, even in cosmology. The *self* in self-organization must in some sense be recognized as an independent agent, with its own self-definition and point of view, distinct from those of human beings.

Generality, predictability, and eternal laws distract from the significance of particulars and the suchness of reality. On the other hand, the very nature of the emergent phenomena on which laws (and our ordinary experience) actually depend is that they are insensitive to the microscopic details of the systems involved.[2] The behavior of large aggregates cannot generally be derived from the behavior of their individual particles, nor can the motions of those particles be reconstructed from the emergent behavior. This is not only because of exponentially increasing computational difficulties (as the number of particles grows), but also because collective organizing principles give rise to qualitatively surprising new properties of large aggregates.[3] The whole is more than the sum of its parts. While this limits reductionism, on the other hand it justifies laws as emerging through nature's own self-organization, and not only for their human utility for prediction and control.

Deterministic laws are designed to reach down (and up) from our intermediate position in the scale of things, in such a way to guarantee control as far as the eye can see and the mind can reach. Nature itself, on the other hand, offers up order "for free" from its own random depths. There is evidence—in some approaches to quantum gravity, for example—that particle physics and even space-time represent emergent collective phenomena. This reverses the traditional hierarchy in which particle physics is the most "fundamental" science, to which complex and collective behavior is reducible.[4] The assumption that fundamental laws are to be found at the smallest scales is a prejudice of this hierarchy: particle physicists take their own investigations to be fundamental, because these deal with what they take (in each generation) to be the most basic parts of matter/energy according to the reductionist program. That the truly fundamental laws might inhere in another scale would entail abandoning reductionism in favor of a more organic holism. Linear determinism would have to be forsaken in favor of laws that operate in multiple directions, implying mutuality in our relationship with nature. Like deductionism generally, the geometrical modeling of time and associated determinism may have actually retarded the development of physics.[5] We must reexamine change from a point of view in which it is fundamental, rather than seeking either unchanging laws or a transcendent meta-time in which laws may vary.

Warp and Woof in the Fabric of Science

In this work, we have explored complementary themes of the scientific process—such as materialism and idealism, empiricism and deductionism, and holism and reductionism. Reduction seeks to explain things in terms of their parts, or of some more basic level of organization. The opposite (or complementary) strategy seeks to explain things in terms of the greater wholes and relationships of which they are a part. Similarly, deductionism seeks to explain patterns in terms of idealized models; its complement must seek to understand the limitations of models in relation to the realities they model. Such complementary approaches form the weave of the scientific process.

The models to which nature is assimilated in classical physics are idealizations that often ignore the organic complexity and interconnectedness of nature. The practical benefit of such modeling has until recently outweighed its infidelity to natural reality. As productive as this approach has been for technology and theoretical coherence, today this approach may stand in the way of further scientific progress. The shift from a science of parts to a science of the whole implies a shift from the notion of transcendent, externally imposed laws to the notion of a universe that orders itself.[6]

The identity of an individual thing is relative to the parts that constitute it as a whole but also to the greater wholes of which it is a part. Since the part cannot be properly described without reference to the whole, there is need for an integrating strategy to serve as a complement to the differentiation of reduction. A real whole can never be fully specified, however, just as part is necessarily a fiction.

A whole involves connections that extend both directions in scale.[7] Its exploration is limited by real epistemic circumstances—such as the horizon of the visible universe, set by the finite speed of light. A similar limit applies at the opposite end of the scale, set by the finite grain of light. Knowledge of an ultimate part or an ultimate whole is conditioned by physical facts that pose a limit on what can be observed in either direction of scale. In both cases, one has the freedom to settle for a product of definition in place of an ambiguous or inaccessible reality. Parts can easily be redefined as elements of a deductive system, but the real whole remains more elusive. We are used to reductionist thinking, but far less adept at its integrating opposite. Such mysteries as inertia, non-locality, and the non-discernibility of particles may be clues to the interconnectedness and holism of a self-creating world.

Another aspect of holism is that analysis cannot be restricted to single causes tailored to render the world predictable. Rather, analysis must be in terms of the joint action of diverse factors, which tends to yield surprises. (Indeed, some of the best scientific discoveries have occurred by accident.) For this reason, at this stage, an organic holism bears more promise of discovery than reductive mechanism. In general, the treatment of nature as a definable object should give way to its treatment as an indefinite presence, capable of responding freely to our provocations in a mutual adaptation. Again, I emphasize that the intention is not to say what nature *is*, let alone to lapse into mysticism, but rather to prescribe a novel approach that may lead to new results.

In very general terms, like any form of cognition, science is mandated to embrace a basically realist attitude, seeking to know *what is*. As with human consciousness generally, this should be tempered by a positivist or skeptical attitude, which seeks to question knowledge and understand its conditions and limitations. We have seen that human cognition in general embodies both these functions in a kind of dialectical balance. The general outward-looking aspect of mind, which holds experience to refer to a real external world, is tempered by the role of subjective consciousness, which challenges the nature and validity of experience in a given moment or context, if only by bracketing it as "phenomenal." These two functions have their counterparts in scientific cognition, where they should also remain in a dialectical balance.[8] Yet the "realist" function tends often to dominate, and positivism or skepticism tends to be viewed as playing only a peripheral role, relegated to certain periods in history. The overweening desire to arrive at a complete and reliable picture of the world may upstage the equal value of bearing in mind the construction of this picture at each stage.[9]

It is one thing to aspire to unify a field of human endeavor, such as physics. It is quite another to hope to capture the totality of nature in a single theory. Let us distinguish between the unification of knowledge and its completion in a unifying theory. The notion of completeness represents a final triumph of mind over matter. It depends on exhaustively reducing physical reality to systems of thought. The more modest goal of unification admits that understanding may never be final, and that research programs can be organized differently. Yet the fact that scientists tend to concern themselves with specialized questions in their own fields means that the broader questions that could unify science—and, indeed, our view of nature—tend to go unexplored.[10]

The program of unification in physics covers not only the fundamental forces and entities of nature but also fundamentally divergent theoretical approaches in realms of the very small and the very large: quantum theory and relativity. In terms of concepts and approaches considered fundamental, the middle ground of the human scale has been relatively neglected. Nevertheless, it may provide the key to uniting the extremes, for both quantum theory and relativity are essentially mechanistic conceptions involving linear causal chains and isolated systems. It is increasingly unclear how such thought can apply to the universe as a whole. In contrast, the "middle ground" is the zone of life and of complex interconnection, involving processes of self-organization and multiple and circular causation. New mathematical tools make it possible now to explore this ground theoretically and try to discern how it might apply to the large and the small.

Self-Organization

Erwin Schrödinger noted that while order emerges within living systems through dedicated programs, it can also emerge within nonliving systems simply from thermal noise.[11] This possibility of the spontaneous emergence of order from apparent chaos has since spawned whole new sciences concerned with chaos, order, and complex systems.

Nineteenth-century physics, dominated by thermodynamics, envisioned a heat death of the universe as the final triumph of inert matter over order maintained within living things. It seemed to be an inevitable outcome of mechanist thought. Only relatively recently has the notion of spontaneous self-organization been employed to explain the existence of organisms in the first place. That is, not only are organisms agents that exploit energy differentials, as human technology does, but such differentials can give rise to "spontaneous" local increases of order that can ultimately take the form of organisms. Moreover, the concept of self-organization provides a common ground for living and nonliving matter.

Could the universe as a whole then be considered a self-organizing system? Normally what one thinks of as self-organizing systems are driven by a flow of energy through them from outside, which presumes an environment. In what sense can the universe be considered to have an environment or to be far from equilibrium? To some extent, we can turn the question on its head by realizing that equilibrium is not a natural state insofar as it can occur only in systems that are artificially distinguished from the whole.

Self-organizing systems are driven by a flow of energy that keeps them out of equilibrium. Gravity may provide such an energy source for large systems like galaxies. To the extent that conditions in the universe at large may once have resembled those in the disks of forming galaxies, it is reasonable to think that the universe as a whole might be self-organizing.[12]

The sensitivity to initial conditions responsible for deterministic chaos has a complement in the notion of "convergent flow," which means an *in*sensitivity to perturbations or initial conditions, a robustness in which divergent states are brought closer together in a mapping of many to one. This is a way of describing the stability of homeostatic systems.[13] While physical reality seems in many ways to exhibit "divergent flow" (sensitivity to initial conditions), we need to look instead for processes that actively converge toward a stable end. To the extent it holds matter to be the passive servant of governing laws, current physics does not do this.

Our very ability to systematically organize experience may pose an obstacle to understanding self-organizing systems. While the organism adapts to changes in its own state, for example, the observer may perceive it as adapting to, and acting upon, the environment as perceived by the observer. The nature of the organism, however, is to be *self*-defining; it is only incidentally an object of human definition and analysis. How this is implicated in the study of nonliving systems is a topic for further study. Though we are not without clues, at this point one can scarcely conceive how the universe might have organized itself, let alone in what sense it may be self-defining. Nevertheless, at the least, artificial divisions imposed on nature should be distinguished from its true parts, which must be the result of self-organization rather than organization imposed by the observer. Only self-organized entities have objectively proper parts—and that may include more than living systems.

A physics of self-organizing systems would by definition not be mechanistic. Perhaps, if not too literally, we must again think of the cosmos in biological terms. "Life" is presently viewed in stark contrast to inert matter. We could underline instead the commonality of living and nonliving matter in terms of concepts of self-organization.

Toward a Second-Order Science

The classical exclusion of teleology from science, with the narrowed focus on efficient cause, has its moral parallel in society: a diminishing sense of individual responsibility. As the institutions of mechanized society become ever more abstract and impersonal, responsibility along with personal choice are constrained. Even in courtrooms, one can now hear pleas of innocence on the grounds of genetic or chemical determinism. As science is at an impasse in its preoccupation with mechanical causation, so is society at an impasse under the sway of the mechanist metaphor. And if science would embrace a broader concept of causation and agency, its exemplary influence might encourage a more responsible notion of selfhood and agency in society.

Can science provide the basis of a new morality, particularly one that will respect nature? It ought to be within the human potential, as well as to our interest, to approach the well-being of the planet rationally and "objectively"—that is, in a manner that takes account of human subjectivity and purposes, and does not merely treat the natural world as an object for use. However, science has a checkered history of reflecting human-centric values aimed at the exploitation and "conquest" of nature. In effect, it has served as the intellectual arm of a reconstruction plan under the occupying human force. If physics and biology are not suitable consultants on the issue of planetary survival, can the social sciences do better? Can they see incisively what is killing a society from within its own institutions?[14]

At least the social sciences include human agency and subjective consciousness, which physical science so far does not. Vico's admonition was to study what we know firsthand: human institutions. That includes the institutions of science, of course. Yet the study of science as a human endeavor has largely been left to philosophers and historians of science, anthropologists, and sociologists. The division between soft and hard sciences has allowed physics to continue as a first-order science, relieved of the need to self-examine. The success of this arrangement is obvious, surrounding us with technological marvels. This does not mean, however, that science could not benefit from self-examination, explicitly including the observer, experimentalist, or theoretician in its formulations.

It is this concern that gives philosophical justification to the distinction between first- and second-order science, originally imported from cybernetics, which involves feedback and circular processes. This distinction does not yet widely appear in fields that are not concerned with such processes or the participatory role of the scientist. These terms must have somewhat modified meanings when applied to physical science. Accordingly, I propose this broad definition: while first-order science studies the object by excluding the subject, second-order science also studies the role of the subject both in formulating and in interacting with the theoretical object.

While physics typically deals with systems from which the observer can stand apart, study of the universe as a whole must include the observer by definition. Both cosmology and theoretical physics are highly speculative sciences, participatory in being theory driven and often dependent on long chains of inference based on indirect evidence. Their speculations can border on metaphysics, and their methods are sometimes controversial, as in the use of anthropic reasoning. They are often concerned with fundamental questions, such as the origin of the universe and the ultimate horizons of knowledge. For such reasons, it may be appropriate to include within these sciences an analysis of the theorist's involvement. Thus, cosmology and theoretical physics are reasonable candidates for a second-order science.

An obvious objection, of course, is that a second-order approach may not be feasible without obstructing the primary aim to study nature. It was to separate effects due to the subject from effects due to the object that science developed as we know it. This strategy is no longer effective in extremes where the observer's involvement cannot be ignored. The exclusive focus on the object has yielded tremendous gains in technology, of course. Yet it may also have yielded a distorted picture of nature. One could therefore say that a "complete" theory or science is not one that maps the object one-to-one but rather one that includes in its map the theorist and the processes of mapping. Just as self-consciousness enhances objectivity, in the long term this sort of completeness would promise a more, not less, objective science.

A sweeping effect of the Scientific Revolution had been to redefine natural philosophy as first-order science. Classical physics remains the paradigm of a first-order approach, in which the physical world, and not physics or the physicist, is the proper object of study. This reflects the outward focus of the organism motivated to take an interest in its environment, whose cognition is part of its strategy for survival. On that primary level, it does not serve the interests of science—any more than the survival interests of the creature—to dwell on its own cognitive strategies or limitations. But this is so only on that level; just as reflexive consciousness obviously plays an essential positive role in human affairs, the question at hand is the possible benefit to science of considering effects on a secondary level. The very fact of self-awareness establishes a reflexive point of view for theorists, who have the opportunity to consider their own activities in the light of such concepts. And, while individual sciences adhere to first-order description, the notion of *science* at large suggests a complex human enterprise, possessed of self-reflection, as much as it suggests a comprehensive body of knowledge.

On the other hand, one of the ways a given discipline can maintain its first-order focus is through the clear delineation of its subject matter in contrast to other specialized disciplines. Yet every inquiry leads naturally beyond its borders. Physics and cosmology are now positioned to struggle with grand questions, such as how the universe as a whole could have arisen and how it will end. To answer such questions scientifically may require more than extending and joining existing bodies of knowledge, and more than a new theory or ontology framed within conventional terms. The grand questions concerning what physically exists (the object) involve how the theoretical observer (the subject) relates to it. For in the case of the universe as a whole, the observer can no longer be presumed to stand outside the system observed. At the other end of the size scale, the measurement problem and discreteness in quantum physics seem also to implicate the observer's or theoretician's participatory role, calling into question the traditional separation of subject and object. The grand question of what exists involves the grand question of how consciousness fits into the picture it pictures.

This book began with a fundamental premise: all cognition involves a conjoint input of subject and object. Given that premise, it is clear that no science can be complete that ignores either. It was also noted that the usual way to explore with clarity the influence of the object upon the subject is to hold the latter fixed. This is one aspect of "controlled" experimentation and the meaning of "objectivity" in physical science: holding in check undue influences of the observer/theoretician/experimentalist in order to establish a consensus regarding what physical phenomena are in their own right. This consensus is largely confirmed through the feedback obtained in successfully manipulating the world in predictable ways—whether in experiment or engineering. However, this leaves open the question of just what influences are "undue."

While consensus has a social value as part of democratic political process, not many scientists would accept the notion that truth is a matter of majority or even unanimous opinion. A further role of consensus is to sidestep debate that would cast doubt on the premise of objective truth behind scientific "fact." But excluding consideration of subjective influence does not eliminate it. Rather, decisions about what sort of influence might be involved are simply not on the table for discussion. Thus, the background effects of idealization, mathematization, deductionism, or the philosophy of mechanism, for example, are not considered. A first-order science is closed to such considerations, its focus on a foreground it defines. Yet their effects may be real, accumulating in the "unconscious" of science until they can no longer be ignored and irrupt in some crisis or paradigm shift.[15] While that may be a workable arrangement in some ways, the crisis tends to be formulated and dealt with in the established first-order terms, while it may point in an entirely different direction.[16] A longer view would systematically embrace and integrate issues potentially arising from the "repression" of the subject factor.

If we look deeper than the pretension of science to objective knowledge, to recognize its social commitment to provide a certain kind of practical empowerment, then it becomes apparent that science forms part of the general management of society. It is the technological extension of human powers, both active and cognitive.[17] The task for a second-order science is to include in its view of natural reality how its own activities shape that view. Hence, a second-order science would view itself as an active player in a larger game. It would recognize itself as an integral part of the direction of society. Its theory making coevolves with society itself, influencing society and influenced by it.[18] It would no longer pretend that its theories do not affect what they are about (for science has had obvious effects upon nature), nor claim an objective description of nature in the traditional sense. It would certainly not pretend that social, economic, and biological systems resemble physical systems; on the contrary, it would embrace the idea that the felicitous approach to physical systems at this point should draw upon the former.

In the same way that subjective consciousness generally empowers human beings, giving them a reflective edge over other forms of intelligent life, a second-order science would be more complete, realistic, and objective than first-order science. It would include knowledge of how first-order theories are to be used.[19] Like reflexivity generally, it bears the potential to free scientific thought from unrecognized assumptions. Science is our modern interface with nature. It should be integral in the social planning intimately involved in that interface. Above all, it should be an advocate for the immanent reality of nature, no longer considered as a background, context, or venue for the human drama, let alone as a backyard to pillage for resources. It would recognize nature as a political force with which to recon and negotiate, as we do with foreign powers with which we wish to trade and keep peace.

A disadvantage of the second-order approach is that claims to traditional notions of objectivity and disinterest must be abandoned. On one level that means little more than giving up privileged claims to social authority presently endorsed by the first-order approach. If we believe that including the subject ultimately enhances objectivity, this fear is more about political clout or social standing than about achieving an unambiguous understanding of nature. Science hardly exists in an ivory tower; scientists and their institutions are constantly jockeying for funds and prestige, sometimes compromising their espoused disinterest. A more serious concern is that reflexivity might defeat the very reason why a first-order approach was established in the first place: to let nature speak for itself, without undue conceptual interference. It might mean that it would be impossible to formulate laws of nature such as we have known them. This, however, simply begs the question at hand; for science *is* human interference in nature. The intervention involved in experimentation and in formulating laws may not be unwarranted, from a human point of view, but its value might be improved by better taking its intrusiveness into account.

Another objection is that inclusion of the subject would entrain circularity, paradox, and logical inconsistency—as occurs generally in problems of self-reference. However, science is replete with circularity and inconsistencies already, generally ignored! Examples include the problem of cognitive domains, the cosmological fallacy,[20] the psychologist's fallacy, the logical inconsistencies involved in extreme conditions (infinities and singularities) and at extremes of scale (the measurement problem, multiverses). In fact, we deal with self-reference normally in everyday life, with mostly positive rather than negative consequences.

Responsibility

As social beings, we are accountable to others even for how we view the world. Ideally, science is a *commons* that belongs to the whole of society and answers to humanity at large, beyond corporate, state, or academic borders. It is humanity's modern tool to understand our place within the universe and to mediate our relationships with it. It is crucial to distinguish what is essential to science as our ambassador to nature from various aspects of its practice, institutional organization, and specific motivations—such as control, prediction, technological progress, national or corporate advantage, etc. A role of second-order science is to critically appreciate the influence of such goals and considerations on shaping knowledge and the ideals and directions of science.

It is understandable that the first scientists redefined natural philosophy as first-order science. They were rebelling against a confusing scholasticism that discriminated little between reason and faith or between evidence and imagination. On the other hand, they were part of a religious reform movement that questioned the ability of reason to overcome human depravity or to know God without drastic spiritual change. Perhaps it seemed, in that light, that reason could at least approach the external world effectively. However, if it could not reliably interpret scripture, why should it reliably interpret the Book of Nature?[21] The question concerns the relationship of science and reason to nature, the scientist's state of consciousness, and the limits of scientific thought. It is still relevant in our era, with implications for our very survival.

The natural philosophers of the Enlightenment resolved their doubts by placing faith in reason, and by adhering to a third-person perspective to the exclusion of subjectivity. They abandoned utopian social projects in favor of creating a science that could physically reform the world through technology. Their successors followed the same detached approach to society as to nature by creating intellectual abstractions, social sciences, and economic institutions divorced from direct moral concern. The relevance of "values" went underground—distantly reflected in such intellectual values as simplicity, unity, and mathematical elegance.[22] Nature itself was seen as a machine to which values are irrelevant. The fact that humanity must be part of that machine seemed to diminish human autonomy, worth, uniqueness, and moral responsibility.[23] Yet acknowledging that determinism and mechanism are but human *ideas* can put responsibility back into human hands. In the end, we alone bear responsibility for our constructs and the world we make. We cannot blame them on reality.

The fact that there are no clear boundaries to epistemic responsibility obliges vigilance. A second-order approach can unwittingly lapse into a first-order approach.[24] A meta-perspective can assume the same overconfidence that characterizes first-order science; it can proceed to objectify its domain in the same way that characterizes naïve realism. In general, the act of stepping outside or beyond any system of thought establishes a new perspective, which our ingrained realism tends nevertheless to present in the usual objectivist terms. These must in turn be relativized, stepped beyond, in a dialectical process.

Post-humanists and science fiction writers imagine a future in which technological man has complete mastery of matter and takes an active hand in the future evolution of life and intelligence. Human beings, or their bionic descendants, would engineer planets and even galaxies. Whether or not such a vision is anything but a masculine delusion of power, it points to the more modest possibility that humanity could take responsibility for its unique position as conscious caretaker of at least *this* planet. One might call this possibility "third-order science."

Toward Natural Rights

We have argued that science itself would benefit from a shift in attitude toward its objects of study. Certainly nature would benefit. Because subject and object are inseparable, this shift involves both. Led by science, society would *treat* nature differently because it *views* it differently. Science and society alike would acknowledge that nature is something different than what we had thought.

To move toward mutuality with nature means shifting from a subject-object relationship to a relationship between subjects: *I/it* becomes *I/thou*. The subject must be receptive as well as imposing. Use must give way to coevolution and respect. This means letting go of centuries of traditional entitlement to treat nature as a provision for human benefit, a raw material for the re-creation of the world in a humanized image. It means viewing nature in terms of mutual relationship, as we do with other human beings and even our pets—with consideration and give and take. To be permeable both ways, however, the relationship cannot merely project human consciousness onto the nonhuman other.

The circle of moral concern has greatly expanded over time. In principle, it has always applied more or less only to members of one's own group. However, membership in this group, based on some criterion of likeness, has gradually grown more inclusive. The axial religions played a key role in this expansion, at a time when cultures were forced into increasing contact and potential conflict. Where once the definition of personhood meant membership in one's tribe, and within that tribe may have included only males of a certain race and class, it now includes in principle all members of our species. For many, moral concern—if not personhood—extends beyond the species boundary to include other creatures and whole ecological systems. The logical extrapolation of this expanding circle would include even the inanimate world.

We shall argue that the object of moral concern must continue to extend beyond human (and, more recently, animal) rights to include nature at large. Religions, which codified moral concerns, evolved to regulate society, serving specifically human interests—notably the interests of tribe or class. To serve those interests, the patriarchal religions and the science that grew out of them demeaned the immanent reality of nature. We now have almost universal laws (whatever the actual practices) concerning human rights. But short-term human interests continue to conflict with longer-term interests tied to the well-being of the planet. Without returning to "nature worship," we must at least accord to nature legal rights. After all, it was literally aboriginal—here before us!

Of course, centrism of any sort can always draw a line to discriminate against the outsider. Even members of what is now recognized as the human biological species are still regularly denied moral concern, legal rights, and personhood. Members of other tribes have often been considered subhuman, literally fair game. We seem ever ready to regress to this kind of in-grouping, so that a universal charter of human rights is a truly grand accomplishment, however poorly it is respected in practice. A next, if radical, step would be a universal charter of the rights of nature. But such a project remains plagued by many questions: Which entities should have "rights" and on what basis? Other primates, cetaceans, elephants, dogs, cats? All vertebrates? All sentient creatures? All life forms? All matter?

Other questions arise, concerning how to administer laws based on such distinctions and enfranchisements. If nature is to be an object of moral concern, even granted legal rights, do these entail the responsibilities normally attending "citizenship"?[25] What would it mean to hold nature accountable to its responsibilities? Such questions could keep many heads spinning for decades. However, nature does not have to be considered a moral *agent* in order to be considered an object of moral concern by those who recognize themselves as moral agents. It does not have to qualify as a human person in order to be accorded the respect due human persons. Natural agency does not have to mean human agency. The natural world can be entitled to legal rights in the human world without being defined as a human player. Nature can be ambiguous without ceding its rights.

For an object to qualify as a subject does not depend only on its own characteristics. That would be to focus only on the object side of our own cognition, placing outside ourselves the burden of justification for the discriminations and schemes of classification we make—the very stance that has been the problem all along. Rather than make natural entities compete to be accorded rights in our world according to how we evaluate their merits, the shift in attitude must come from the subject side. It is not a question of which things deserve our respect, but of how to summon that respect within ourselves. It is not "our" world into which "they" must fit, but a common world in which we must try to coexist.

Perhaps because of a long tradition in which the reign of law regulates human conduct, often in place of immediate moral sentiment and empathy, we moderns are surprisingly open to legalistic arrangements such as treaties, international trade agreements, and even aboriginal land settlements.[26] In practice, we often seem motivated by respect for law more than by the actual condition of other people, other creatures, or "the environment." Perhaps this reflects the general need for a social contract to mediate human relationships. One thinker has extended this notion to include a *natural contract* in which we would "set aside mastery and possession in favor of admiring attention, reciprocity, contemplation, and respect."[27] Perhaps modern legalism also reflects our essence as the self-defining creature, and the compulsion to establish a world defined by human intentionality, read as law. In any case, it opens a plausible avenue through which to protect nature—indeed, to secure its rights as an inalienable other.[28] Though the personhood of nature is far from a scientific hypothesis, it is interesting to imagine what its scientific consequences could be as a research program.

Sadly, the accomplishments, dreams, and real prospects of space exploration may serve to dampen enthusiasm for the preservation of *this* planet. Despite the emotional impact of the first images of Earth as seen from space, the possibility of commercial space travel may have the longer-term effect of reinforcing the idea of Earth as a disposable stepping stone in a larger manifest destiny. There are, after all, innumerable other planets out there to pillage! Might we not ultimately live in self-contained artificial environments while migrating to the stars? The wide-open universe represents a new frontier and a new playing field, just as the New World did for early modern Europe. Science fiction nourishes the fantasy that our species has outgrown its habitat of origin, just as individuals outgrow their dependency on parents and leave home.

This fantasy demonstrates the hazard of thinking purely in "objective" terms—that is, in terms of the uses of things and the possibilities they afford—especially in terms of isolated systems or in terms of inert materials as opposed to living matter. As engineers, we have little essential bond to this planet of our origin. Rather, in the tradition of our religions, it is easy to regard ourselves as entitled to do with a vast universe as we please, following our appetites as our forebears did with terrestrial "resources." Though we are products of nature, it appears to many that our destiny is to escape the bonds of gravity and even embodiment.

Of course, I believe that such fantasies, however seductive, are literally unrealistic. Moreover, our dual existence as subject and as object involves a two-faced relation to Mother Nature. On the one hand, she nourished us and made our freedom possible; on the other, she limits us and is an embarrassment to us in our adolescence. Post-humanist dreams are based on understandable delusions of power and grandeur, hopeful extrapolations of accomplishments to date. In place of this schizoid vision of nature and quest for unlimited power, I propose that we regard the natural world as an *other*—inscrutable but perhaps lovable—neither "mother" nature nor the creation of a patriarchal god. I propose that we regard ourselves, rather than nature, as in need of reform. This attitude cuts across established categories, to apply equally to the depths of space and to the rain forest. There is then no category of material existence exempt from moral concern, no second-class citizens of the universe.

Notes

1. Toulmin, Stephen. 1982. *The Return to Cosmology: Postmodern Science and the Theology of Nature.* Berkeley: University of California Press, p. 237.
2. Laughlin, R. B., and David Pines. January 2000. "The Theory of Everything." *Proceedings of the National Academy of Sciences* 97:29.
3. Laughlin, R. B., D. Pines, J. Schmalian, B. P. Stojkovic, and P. Wolynes. January 2000. "The Middle Way." *Proceedings of the National Academy of Sciences* 97:32. The authors focus on the properties of matter at the "mesoscopic" scale, the size order of large organic molecules, but examine the properties of nonorganic matter at this scale as well, finding evidence of organizational principles in glasses, for example.
4. Smolin, Lee. 2006. *The Trouble with Physics.* Boston, New York: Houghton Mifflin, p. 315.
5. Cahill, Reginald. 2003. "Process Physics." *Process Studies* supplement, p. 19.
6. Smolin, Lee. 1997. *The Life of the Cosmos.* Oxford: Oxford University Press, p. 15.
7. Bohr held that quantum phenomena are characterized by "wholeness"—meaning that they cannot be further analyzed because of the indivisible quantum of action. The focus then was on the bottom end of the scale; the focus now must turn to the indivisibility of large-scale phenomena.
8. Cf. Holton, Gerald. 1996. *Thematic Origins of Scientific Thought.* Cambridge, MA: Harvard University Press, p. 393.
9. Hence, astronomers may glibly discuss highly theory-dependent estimations (of the age of the universe, for example) as indisputably precise, though involving interpretation and wide margins of possible error in the "ladder" of distance scales upon which they are based.
10. Toulmin, *The Return to Cosmology*, p. 230.
11. Schrödinger, Erwin. 1944. *What Is Life?.* Cambridge, UK: Cambridge University Press, chapter 7.
12. Smolin, Lee. May 1995. "Cosmology as a Problem in Critical Phenomena." arXiv:gr-qc/9505022v1.

13. Cf. Kauffman, Stuart. 1995. "Order for Free." http://edge.org/documents/ThirdCulture/zd-Ch.20.html.
14. Ernest Becker, Ernest. 1971. *The Birth and Death of Meaning*, 2nd ed. New York: Free Press, p. 159.
15. The cycle of "normal science" and "scientific revolution" described by Thomas Kuhn in Kuhn, Thomas. 1962. *The Structure of Scientific Revolutions*. Chicago: University of Chicago Press.
16. One sees parallels in "crisis management" within economics and in "coping mechanisms" on a personal level.
17. It is not sufficient simply to accept the empowering aspect of science as its "true" nature because human beings, their activities, and their creations have an indeterminate nature. We are the creature that self-defines, and we are at liberty to redefine science.
18. Umpleby, Stuart. 2014. "Second-Order Science: Logic, Strategies, Methods." Washington, DC: George Washington University. www.gwu.edu/~umpleby/papers/wmsci_sos3.ppt.
19. Umpleby, "Second-Order Science."
20. Smolin, Lee. 2013. *Time Reborn: From Crisis in Physics to the Future of the Universe*. Toronto: Alfred A. Knopf Canada. His term for applying concepts outside the realm in which they are formulated.
21. Menuge, Angus. June 2003. "Interpreting the Book of Nature." *Perspectives on Science and Christian Faith* 55:88–98.
22. Maxwell, Nicholas. 2011. "Does Science Provide Us with the Methodological Key to Wisdom?" PhilPapers. http://philpapers.org/rec/MAXDSP.
23. Earman, John. 1986. *A Primer on Determinism*. Dordrecht, NL: D. Reidel, p. 250.
24. Kenny, Vincent. March 2009. "'There's Nothing Like the Real Thing': Revisiting the Need for a Third-Order Cybernetics." *Constructivist Foundations* 4:100. www.univie.ac.at/constructivism/journal/.
25. Duguid, Stephen. 2010. *Nature in Modernity: Servant, Citizen, Queen, or Comrade*. Pieterlen, Switzerland: Peter Lang, p. 282.
26. For example, in Canada.
27. Serres, Michel. 1995. *The Natural Contract*. Ann Arbor: University of Michigan Press, quoted in Duguid, *Nature in Modernity*, p. 284.
28. The first ever such law was passed in Bolivia in 2010, the "Law of the Rights of Mother Earth."

20

The Stance of Unknowing

"Science is a quest to convince yourself and others of something you only guess to be true."
—Michael Brooks[1]

"You know something's happening, but you don't know what it is, do you, Mr. Jones?"
—Bob Dylan

The Need to Know

The feeling of security—if not its reality—lies in whatever certainties we can conjure. Technology shelters us from nature's vagaries, and science from its ambiguities, while other cultural institutions and practices shelter us from psychological anxiety, and the state shelters us from the indifference of other people. Yet the very fact of subjectivity places human beings in a position of chronic uncertainty, for how can we know that knowledge and perception are true to reality? This problem has preoccupied western philosophy from its inception in Greek thought: how to separate the objective from the subjective and justify knowledge. But it is a challenge that is deeply motivated psychologically. The need to *justify* such inductive leaps as Hume criticized, for example, simply reflects the insecurity that motivates them in the first place.

Survival depends upon decisive action, based upon good information, and action requires a degree of certainty in the face of doubt. Species without statistically accurate knowledge may disappear. The individual, however, can never be directly assured of the quality of such information in immediate situations, however intuitive it seems or upheld by tradition. Self-awareness entrains doubt, since one always deals with one's own states of belief and those of others. One way or another, life is ultimately based on faith. While faith has always been the prerogative of religion, it has an inescapable place in science too.

A chief strategy in the campaign against uncertainty has always been to create familiar constructs to stand in the place of the unknown. We have made a whole civilization for this purpose, consisting of rules, customs, buildings, streets, artifacts, and ideas, which constitute our actual milieu. Even mathematics is the art of formally manipulating unknowns as though they were known. Scientific theory stands in for nature itself. Experiments are contrived versions of the natural phenomena they investigate. Knowledge is dominated by an idealist component to the extent that we prefer the certainties of deductive systems to the contingencies of the found world. As pointed out four centuries ago by Vico, only human constructs offer that kind of certainty. As a side effect of this drive toward certainty, understanding is channeled toward artificial situations and the immanent reality of nature is often eclipsed by ersatz versions of it.

The Dualism of Found and Made

Given the inevitable dualism that attends self-consciousness, the quest for certainty casts its own moving shadow. The duality of the found and the made has variants in such polarities as ambiguity versus clarity, opinion versus truth, probability versus determinism, empiricism versus logical certainty, faith versus fact. There is, of course, a middle ground between logical certainty and radical skepticism concerning empirical knowledge. This is the ground claimed by science.

Christianity rejected the Greek tradition of unbounded speculation, which was attended by accountability for one's views as personal claims and a moral association of words with the character of the speaker. In contrast, God spoke through his prophets, preachers, and scriptures, making personal opinion irrelevant when not blasphemous. Thus, mere philosophy or speculative thought was rejected in favor of unconditional faith or adherence to received doctrine.[2] Christianity eventually went a step further by making speculation dangerously heretical. This forfeiture of responsibility for one's subjectivity reflects the influence of Plato in positing a truth that exists a priori, eternally, and without reference to the speaker or the testimony of the senses. It may reflect also the influence of written text as an apparently objective authority. Subjectivity was partially reclaimed by society in the early modern period, which also saw the rise of first-order science, from which subjectivity was systematically banished. In some sense it is unfinished business for science, for religion, and for the society in which both play such a key role to reclaim subjectivity within their domains. In science, that would mean taking greater responsibility for the foundational assumptions on which it is based and upon which society trades. And prime among those assumptions is the tacit belief that nature is ultimately reducible to human thought.

Functions and Dysfunctions of Positive Knowledge

Nineteenth-century philosopher Auguste Comte advanced the concept of positive knowledge in order to replace mere belief with rational thought. In his view, theology, faith, and metaphysics should be superseded by science based on careful experiment and on observation of natural processes. Positive knowledge is thus carefully crafted and vetted information, especially knowledge of the physical world useful for human progress.

The quest for positive knowledge stems from the deepest layers of our conditioning, which compel us to seek the imprimatur of the real for any ideal we hold dear. We want to be angels, not brutes—free spirits, not prisoners of genes and gravitation—and for reasons beyond mere preference. We seek sanction for these aspirations in our understanding of the natural world. But nature does not willingly affirm our values or support a special place for us in the order of things. On the contrary, it appears we can only have our way by forcing nature to our will and re-creating it to our fancy.

The qualification *positive* has a broader connotation than a particular notion of what is real, true, reliable, or useful. It reflects an attitude or approach to experience in general, quite apart from what is experienced. What exists, or is fact, is a different question than how to approach it. Yet the scientific worldview has built into it a set of demeaning attitudes toward "subjective" experience in general. It does more than assert what exists. It tells us how to relate to it.

The attitudes or values associated with positive knowledge include certainty, prediction, control, utility, objectivity, and detachment of subject from object. The knower is unaffected by the known and does not affect it. The observer is outside and independent of the system observed—which is inert, neutral, passive, simply *there* to be discovered and used by human beings. On the basis of this approach, science has achieved enormous practical success through technology. It has become the model of reliable, useful, "objective" knowledge. Yet one is free to ask, might there be value in a different and complementary approach, perhaps a different kind of knowledge?

Most of what is considered positive in our world happens to be associated with males—call it the unbalanced masculine principle. This is another way to say that, despite feminism and much social progress, we live in a patriarchal world dominated by certain values and attitudes: power, success, status, control, expansion, action, force, hardness, confidence, certainty, competition, vindication, etc. These values are so universal in the modern world that they are hardly noticed as "masculine" values. The other side of this coin is that complementary values are overshadowed. These include: softness, acquiescence, humility, ambiguity, tolerance, emotionality, caring, nurturance, cooperation, patience, inclusiveness, equality, consensus, etc. Obviously, these qualities are not absent in society, which could not function without them. But for the most part they are scarcely vaunted as ideals.

The quest for power and status leads to extreme disparities of wealth and social well-being. Unlimited growth leads to collapse of the earth as a living system. Overconfidence in economics and technology leads to fatuous dreams of mastery over nature, to loss of self-reliance, and to increased vulnerability of society to natural disasters and other catastrophes induced by human folly. These are ironic side effects of masculine values and positive knowledge accumulated at the expense of wisdom and health.

Apart from the ideal of truth, we see and categorize the world in ways that help us get on. We know what we need to know and ignore what we don't. Even in science, we choose what patterns to look for in data, and have preferences regarding what structure is relevant to our interests, as opposed to what constitutes "noise" or "error." If the world were not inherently ambiguous there would be little need for such preferences. Indeed, the only way that the world could be *un*ambiguous is if it were a preconceived product of definition—that is, not real.

On a very human level, we are assailed by doubts and barely tolerate the passivity of not knowing, of being powerless in the face of uncertainty. We fear being at the mercy of things we don't understand—whether natural or man-made—and know that there is little of which we can be sure except "truths" that turn out to be, in the last analysis, tautologies or mere conventions. Only if life is effectively a story and the world a narrative can one be assured of meaning and a meaningful place within the scheme of things. While this wishful thought is the basic need reflected in religion, as in culture at large, it is also a driving force behind science as a secular narrative that promises a similar assurance: that reality can be assimilated to human constructs.

Science and technology tend to function for modern people in the way that religion did for earlier centuries, as a reassurance that our situation is not untenable. Yet at the beginning of the Scientific Revolution, when science was hardly differentiated from religion, a writer such as Pascal could poignantly take exception to the ebullient promise of positive knowledge. We read between the frank lines of his *Pensées* the struggle between the aspect of thought committed to unequivocal order, represented by the divine, and the aspect that reluctantly admits the unknown, represented by the chaos of nature. Indeed, as acknowledge by Hume, the more one is assailed by uncontrollable events in life, the greater the drive to find order and coherence in those events.

Yet the nature of the object is to remain ambiguous to the subject. One is free to project definitions and determinism upon nature, to imagine that simple algorithms drive the universe. It is a strategy that happens to work, to a large extent, for select human purposes. Apart from the question of *why* it works, the really interesting thing is our apparent need to know what the universe is *in itself*, apart from how such strategies serve us. We are caught between two commitments: survival and truth.

The Stance of Unknowing

The only certainty may be that there *are* no certainties. Despite the appeals of positive knowledge, we control very little. Life—reality—can always surprise us. This alone should tell us there is room for an approach that accepts uncertainty as the natural condition and that embraces it as a state of mind and an approach to life. I shall call this approach *the stance of unknowing*.

It is not a purely negative condition to be lamented but has benefits. Our inclination and natural conditioning is toward positive knowledge, clarity, and control. To a point, reliable knowledge does serve survival. Yet the quest for it can also be counterproductive; it can even threaten survival if it is not tempered by something else. For knowledge that faces outward, by its very nature as a tool, excludes the intentions behind its use and the purposes it might serve, which are simply taken for granted. We need more than to understand the world, as though the world did not include our motivations and ways of understanding it. We need also to understand what drives us—where that comes from and where it may lead. *That* knowledge borders on wisdom. And yet that is the very kind of knowledge traditionally excluded by first-order science and undervalued by society.[3]

To allow such knowledge, one must step back from the modern scientific worldview, with its outward focus, to gain a bigger and perhaps fuzzier view—not because the scientific worldview is wrong, but because it does not embrace the whole of human reality. Science as it stands cannot bring us wisdom, which involves not only the world but also our attitude toward it and each other. Should we then turn to religion for what is missing? That depends crucially on how one understands religion. *Theology*, we have seen, is the counterpart of scientific *theory*, insofar as both assert positive ideas about what exists. Both aim at certainty and downplay human agency in creating knowledge. Neither theology nor theory can bring us wisdom because neither frees us from the compulsion to know; neither fosters a complement to those values of domination that are destroying the world.

This is hardly surprising, in view of the fact that both science and the world's principal religions were created largely by men in the era of social evolution known as patriarchy, which is very much about domination. Of course, certainty, power, and control may now appeal to women as well as to men, given the current prevalence of such values in society and the long exclusion of women from participation in male-dominated games. Yet there is reason to hope that there must exist underground, so to speak, recessive traits of a different kind, whether these are borne by females or males. These represent another mentality—call it the unrealized feminine principle. Rightly or wrongly associated with women, there are alternative ways of being that can serve us individually and collectively if we know how to access them and are willing. The main obstacle, of course, is that they are not currently prized.

I am not proposing a mystical concept, which no one should even try to rationally understand. I do not advocate abandoning reason and positive knowledge, only that they should be tempered by other qualities. Yet it would be obviously paradoxical to try to reduce the stance of unknowing (whatever that is) to known terms. Language and thought are oriented toward definable things, so talking about indefinable things is unavoidably tricky. Yet one can approach them indirectly, just as black holes can be detected indirectly through their influence on visible things, even though they are not visible themselves. The unknown is the invisible complement of the known, and the stance of unknowing is the needed complement of the search for positive knowledge. If we identify the characteristics of positive knowledge and its quest, then we can know something of its opposite as well.

I call it a *stance* first of all to evoke responsibility.[4] Knowledge involves a knowing agent, and so does any effort to bracket knowledge or suspend belief. Someone *stands* somewhere, both in regard to knowing and unknowing. What are the characteristics of this "stance"? First of all, to suspend what is already known or believed, and to resist the temptation to prematurely reassert positive knowledge. To identify knowledge as belief means to disengage from it enough to not be immersed in it. One must step back from the obvious desirability of certitude in order to see it as *the biological need of an organism*. For the paradox of belief is that it appears not as belief when one is possessed by it but as truth. While stepping back from "truth" means suspending one's beliefs, it does not necessarily mean abandoning them. The stance is a provisional measure, a voluntary act, an experiment. It poses the question: "What will happen if I set aside what I think I know?" If that truly happens, the answer cannot be predicted.

This suspension or bracketing of knowledge and belief creates a void—for the purpose of observing what (if anything) comes in to fill it. Without this emptiness, one simply remains immersed in one's current positive knowledge, which displaces other possibilities. If one has complete confidence in that knowledge, then there is little reason to perform this exercise. Hence, a prerequisite is doubt. One must doubt, and even doubt one's doubt! Such skepticism is challenging in an accelerating society based on decisive action, which relies on positive knowledge. Of course, there are real emergencies and things that must be dealt with in a timely manner. But there are also false emergencies and decisions made with unnecessary haste. We live in an age of shortsighted goals and pressured decision-making—which looks more and more like the very lack of wisdom. There is a valid place for doubt, and for remaining in a phase of suspension for as long as it takes to come to a wise decision. There is a place for the lengthy debate required for consensus, including debate within oneself.[5]

Obviously, the increase of positive knowledge is part of our image of progress. Yet the relentless accumulation of data and ideas may in another light be viewed as a kind of pollution, just as the noise of the city drowns out silence and city lights blot out the starry night. Technological civilization, both physically and ideologically, tends to obscure the reality of nature as a force independent of us, which will always remain partially unknown. It tends to exclude any attitude but that of mastery and domination, in search of certainty, security, and control.

By definition, the unknown is nothing the mind can wrap around, express, formalize, control, or master. Like death, ultimately *it* is master. It cannot be reified or made the object of study, and cannot even be named. It lies beyond the scope of definition. It is not the definable *it* but the truly ineluctable *other*. While it has nevertheless been given names (Tao, Brahman, World-in-Itself, God, Great Spirit, Mystery, etc.), those who dare to go that far often take care to go no further. Despite Old Testament warnings, "God" is an image we have graven in our imaginations, to stand as an idol of the mind between us and the unknown, upon the face of which one cannot gaze. The modern counterpart of this image is provided by scientific theory.

The obsession with control leads us to make definite nouns of things sometimes better left as nebulous relationships, denoted by more humble parts of speech. Concepts like "soul" and "God" are not left free to refer to qualities of being, or attitudes, but are pinned to metaphysical entities and propositions. *Worship* and even *contemplation* are understood as transitive verbs, and theology is the "science" of positive knowledge about spiritual things. Even a more ambiguous "spirituality" is usually backed up by definite beliefs about spiritual "reality." On the other hand, scientific theory is the "theology" of nature, a dogma concerning the material world, whose ultimate realities are virtually invisible. The visible and invisible worlds alike are too much with us!

At heart, the religious impulse is an attitude of wonder, humility, and openness toward the mystery of being. This suggests religion as a needed counterfoil to our secular culture of rationality. Yet as we have seen, the origin of science lies *in* religion. The dogmas and histories of religions show that there, too, the need for certainty, closure, and control has reigned. Many believers hold their faith to be a matter of unquestionable truth. On the other hand, the reflective scientist admits that scientific knowledge is probabilistic, ultimately a matter of faith. In the end, the same usurious attitudes, regardless of source or justification, have resulted in the domination of nature and society alike.

The great commitment of Western civilization to positive knowledge leaves it in chronic imbalance. This is a gender issue since it is men who have dominated cultural production generally and the pursuit of knowledge in particular. What is lost in our masterful masculine civilization is the ability to *not* know, to be tentative, to do nothing, to listen, to receive, to open, to surrender, to embrace uncertainty, to leave the wild untamed. Significantly, we have no word for this ability, which I have dubbed the stance of unknowing. It is a capacity to live intransitively—so to speak—without objects for our intentions. While useful, this "skill" cannot be made to serve a predetermined end. The unknown cannot be caged, packaged, or domesticated. Despite religion, it cannot be made a thing to worship; despite science, it cannot be made a thing to manipulate, even in thought.

While it cannot be cultivated, mastered, objectified, or pursued, yet unknowing can be valued. This demands a different attitude. To exercise it, one must do nothing, *not* move, not "approach" at all. In our culture of doing, such passive receptivity is inevitably perceived as a negation, a letting go, a relaxation of vigilant intent—perhaps a weakness. Applied to nature, it means allowing the natural world to move of its own accord, to exist outside the bounds of controlling thought. However, the very autonomy of nature poses a threat to the dominating mind just as it does to the vulnerable and dependent body. The reason for denying the immanent reality of nature is to affirm "masculine" control. In the words of Margaret Cavendish, early feminist critic of science, this is why the learned "are so much afraid of self-motion, as they will rather maintain absurdities and errors, than allow any other self-motion in Nature, but what is in themselves: for they would feign be above Nature, and petty Gods, if they could but make themselves Infinite."[6] In the dualism of subject and object, we have made ourselves active and made nature passive. It is time to reverse this polarity.

Having eliminated the traditional alternatives to our global monoculture, we are condemned to our apparent successes, whose undesired consequences become ever more apparent. How to find a vision of a viable future? The unknown is an untapped resource unconditionally available. It is self-renewing. It cannot be captured, cultivated, or pursued but always escapes like water through any net. Yet, like water, it surrounds us, permeates us, and can support us when we lose our footing. In that sense, there is no "stance" of unknowing, for there is ultimately nowhere secure to stand. An ontology of the unknown is a contradiction in terms; one must focus rather on *attitudes* toward knowledge and certainty. Surveys of the limits of knowledge are important, but so are the social and scientific attitudes toward knowledge and uncertainty in a given age or culture.

A basic premise of this book has been that there will always be an unknown. Science must be understood as a biological strategy of coping with it, not as a surrogate for it. Our evolutionary history may well have channeled thought in certain directions and away from others.[7] The reason why the mystery of the world is ultimately not reducible to human concepts is not because it is alien to us, but precisely because we live immersed in it, as part and product of its reality.

Some examples of things we do not know include: the future (or indeed, in a certain sense, the past); the experience of other persons and creatures; what awaits beyond death; the precise time of our own death or what the weather will be like on a given day in one hundred years; why there is anything at all; what existed before the big bang or lies beyond the visible universe; the true story of what is "going on" here in "life"; whether experience is "real" or a "dream." For none of these examples is there any shortage of proposed answers. It is human nature to speculate and embellish and not to leave questions unanswered, spaces not filled, and fears not assuaged—even if the bewildering variety of responses only aggravates our doubts. On the other hand, these are not things that most people think about most of the time. Daily life itself is the pragmatic answer to such "pointless" questions. I give them here as examples of things about which we are generally unsure when we do bother to think of them. They are unnerving questions, precisely because we know that we cannot know the answers with finality.

The busyness and chatter of everyday life tend to displace wonder and deep inquiry, shielding the potential inquirer within the terms of the familiar, the human and social realms. The compulsive need for positive knowledge is part of the appropriation of nature—that is, of the unknown. It amounts to the assumption that it is better to have a false, partial, misleading, or superficial story to account for experience than none at all. Toward that end, one defers to experts and authorities as sources of reliable stories. Given the ethos of positive knowledge, it may seem irrational, perhaps suicidal, to suspend all claim to certainty.

Yet once the value of the stance of unknowing is recognized, one can embrace it only by suspending the usual compulsions. One must bracket received wisdom, familiar categories and answers, common sense, and self-evident truths. The stance of unknowing is a willing suspension of belief in the known, for the sake of opening to whatever else there is. This is so even in the realm of science. Especially now that scientists rarely work in isolation, it is all the more important to be able to step outside the consensual wisdom, to suspend belief in the hard-won advances of the latest theory—not to deny those advances, but to keep the path open to further progress.

Science is the successor to magic. The problem with magic was not that it was unrealistic, for it *is* realistic in the terms of the magical thinker! The problem is rather the attitude of mastery, shared by science, technology, and even religion, which renders them alike developments of magical thinking. Science has learned more effective ways than magic to translate thought into reality, justifying its incantations as realistic within a world conceived as material. It is good at sorting out what actually works in such a world from mere wishful thinking. What makes science superior to metaphysics, magic, and religion is its systematic method, grounded in empirical experience through a prescribed cycle of interaction with the world. Yet we have seen to what extent science too is full of unquestioned notions, idealistic premises, self-affirming hypotheses, and ideas beyond experimental confirmation. Along with the intent of mastery, it is just such aspects that reveal its continuity with earlier forms of thought.

The depersonalization of nature facilitated the transition from magic to what we now call technology. One no longer invokes *spiritual* powers to control matter. Whereas nature had been a divine creation and protectorate, in the post-medieval period it was bequeathed to man as God retired from the world. While material reality became recognized as a legitimized focus of attention, this was at the cost of the respect that had formerly been accorded divine creations. It was also at the continued price of nature's autonomy, its existence independent of human interest. Matter was freed from divine will, only to be subjugated to human will.

We have seen that the transition from the medieval period to the early modern was characterized by a shift from allegorical to more literal thought. Chemistry disengaged from alchemy in the shift from a spiritual goal and framework to a materialist framework and an interest in the literal properties and practical uses of materials. The whole course of Western civilization since then can be viewed as an answer to Jesus' rhetorical question, slightly paraphrased: "What does it profit mankind to gain the whole world and lose its own soul?" Modernity is the ultimate Mephistophelian bargain, in which spiritual transformation is traded for the power to transform the physical world. The name of that power is science.

Notes

1. Brooks, Michael. 2012. *Free Radicals: The Secret Anarchy of Scienc.* New York: Overlook, p. 73.
2. Freeman, Charles. 2005. *The Closing of the Western Mind: The Rise of Faith and the Fall of Reason.* New York: Vintage, p. 314.
3. Even psychology, as a first-order science, does not reflexively consider the implications for the science itself of its discoveries concerning human motivation.

4. The term is modeled on philosopher Daniel Dennett's "intentional stance," which means an approach to natural processes *as though* they had been intentionally designed. The intentional stance is a way of looking at the organization of nature from an engineering point of view.
5. The compromise of majority rule is an expedient that can leave 49 percent of the people dissatisfied!
6. *Observations* 114, quoted in Bazeley, Deborah Taylor. 1990. "An Early Challenge to the Precepts and Practices of Modern Science: the Fusion of Fact, Fiction, and Feminism in the Works of Margaret Cavendish, Duchess of Newcastle (1623–1673)," Dissertation. San Diego: University of California, San Diego. www.she-philosopher.com/library.html. Cavendish complained also that the new scientists would not admit that man is both part of nature (and so, not omniscient) and biased in his own limited and partial point of view. She accuses the natural philosophers of an androcentrism that projects (male) human concepts into nature (*Philosophical Letters* 152). For her, nature was not the masculine God's text but an author in her own right.
7. Hamming, Richard W. February 1980. "The Unreasonable Effectiveness of Mathematics." *American Mathematical Monthly* 87:81–90.

Index

abstraction (process of), 14–16, 124
adequacy (vs. accuracy), 133
aether, 170
agency, 241–242, 243, 245–249, 250, 276
alchemy, alchemists, 46, 171, 176, 303
algebra (abstract), xi
algorithm, 23, 64, 66, 74, 77n20, 111, 113, 162, 222, 230, 233–235, 237n5, 248, 267, 297
  algorithmic compression, 222
allegorical interpretation, 157, 158
ambiguity, 70, 81, 294, 296
  of consciousness, x–xi, 103, 105, 138n17
  of nature, 12, 16, 184, 188
analog (vs. digital), 85, 233
analytical (vs. synthetic), 64, 183
ancestral environment, 121
animal heritage, 122, 178
animism, 165, 245
anthropic principle, 89, 271, 283
antifeminism, 123
argument from design, 250–253
Aristotle, 15, 20, 22, 24, 25, 44, 45, 61, 115, 123, 159, 162, 165, 167–170, 201, 205, 219, 240, 241, 248, 255, 260
artifact(conceptual), xv, 86, 122, 142
artificial intelligence, 114, 125, 234–235, 256
artificial organism, 224
aseity, xx
Augustine, 123, 167, 174
autonomy
  of nature, xv–xvi, xvii, 19–23, 61, 103, 129, 130, 166, 301
  of organism, 240, 243, 245, 254
autopoiesis, 38n4, 246, 257n6, 258n27
awareness (vs. consciousness), 4, 8, 17n6, 102
axiomatic system, 140, 179, 184, 189, 200, 256

Bacon, Francis, 3, 45, 46, 125, 174, 175, 177, 186
Bacon, Roger, 173
bible, 44, 61, 155–156, 158, 160–161, 173
big bang, 37, 133, 268, 302
billiard ball, 13, 210, 214
black box, 135–136, 242
black hole
  as fundamental particle, 36, 94–95
  selection theory (Lee Smolin), 266
block universe, 87, 122
body (matter), 4, 6–8, 20, 44, 52, 111, 113, 149, 175, 197n18, 219, 224, 243–244, 247
  human, 52, 103, 113, 119–120, 124, 243, 253–256
  as prison, 119
Bohr, Niels, 7, 135, 138n15, 183, 193, 195, 221, 291n7
Book of Nature, 24, 56n12, 155–163, 173, 185, 204, 287, 292n21

Calvinism, 174
Capitalism, 126, 178
Cartesian theater, 112
category (logical), 27, 94, 147
cause
  as connection, 111–113, 206, 231, 242, 266, 280
  circular (feedback), 68, 73, 84, 242, 270, 282
  efficient, 22, 24, 30, 43, 44, 124, 168, 169, 199, 219, 240, 241, 242, 247, 250, 255, 281
  final, 45, 169, 260
  first, 241–244
  multiple, 73
causality (meaning of), 28, 44, 54, 60, 65, 115, 120, 161, 168

Cavendish, Margaret, 163n5, 170, 180n15, 181n37, 301, 304n6
celibacy, 123
certainty, xviii, 28, 122, 129–130
chaos (deterministic), 50, 67, 68, 75, 202, 281
chunking (coarse graining), 202, 234–235
circular reasoning, 28, 87, 91, 132, 232
classical physics, xvi, 6–7, 28, 35, 62, 68, 88, 107, 130, 171, 206–207, 220, 278, 283
  vs. quantum, 6–7
clock (and ruler), 87, 210, 211
closed system (reversible), 161, 207, 211–212, 262
closure (logical), 15, 84, 142ff, 148, 300
coevolution
  of species, 256
  of subject and object, xi
cognition
  science as, 26, 35, 47, 51, 143, 279
cognitive domain, 132–133, 286
collapse (of the wave function), 95, 208
color (as phenomenal quality), 10–11, 50
commons (public), 173, 286
communication, 92, 110–112, 184
communism, 126, 179
completeness (theoretical), ix, xvii, 7, 26, 33, 193ff, 197n21, 229–230, 262ff, 279, 283
  Einstein and, 193, 221–222, 263
  mathematical, 77n21, 82
complexity, 74–76, 250–253
computable number, 194, 234
computation (digital), xii, 67, 92, 95, 147, 189, 223, 230
computational metaphor, 105, 239
conceptual space, 52, 53, 86–88, 264
configuration space, 86, 122
connection (causal vs. intentional), 111, 113
connectivity 69, 114–115, 207, 254
  intentional, 111–113, 245
  logical, 81, 113
constant
  of integration, 191
  of nature, 60, 184, 191, 194, 196n14, 222, 226, 263, 270
  varying, 263–264
contingency, 16, 50, 122, 166, 172–173, 191, 220
continuum
  vs. discreteness, 37
  space-time, 87
control (attitude of), xviii, 31, 275, 295ff

Copenhagen interpretation, 221
cosmic natural selection, 260, 264–267
counterfactual definiteness, 13–14, 28
creationism, 20, 23–24
crisis (in science), 261
cyberspace, 123, 236
cycle (dialectical), 10, 149

dark energy, 27
dark matter, 27
Darwinism, 47
Death, 131, 150, 280
deductionism, xv, 61, 160, 183–197, 204, 263, 277, 278, 284
deductive system, 15, 16, 26, 63, 82, 144, 156, 161, 183, 184, 186, 187, 188, 189, 190, 191, 193, 194, 200, 203, 207, 210, 261
definition (product of), 24, 53, 70, 85, 93, 141, 159, 161, 189, 230, 252, 254
Deism, 167ff
demon
  Descartes's, 8, 130, 232
  Laplace's, 130–132
  Maxwell's, 130–132, 137n6
depersonalization, xviii, 84, 118, 303
Descartes, René 7ff, 46, 51–52, 171, 173, 193, 195, 223–224, 232, 243, 253
design
  argument from, 250–253
  of nature, 21
  stance, 187, 196n10, 251
detection event, 90
determinism (vs. free will), 59ff, 103, 243–244, 248–249
  deterministic system, xv, 59–60, 62–63, 66–69, 103, 161, 186, 194, 243, 249
digital (vs. analog), 85, 233
discernibility, 214
discrete(ness) 34, 67, 89ff
  definitional vs. natural, 36, 89, 202, 207
divination, 61, 62
DNA
  junk, 237
domination
  by nature, 121
  of nature, 121–122
  of women, xviii
dominion (biblical), 166, 175
Doomsday argument, 233, 226n14
doubt, 70, 117, 130, 145, 284, 287, 293, 296
  Cartesian, 7–8
  positive role of, 297ff

306

Index

dualism (of mind and matter), xvii, 20, 27, 54, 103, 199, 241–242, 247, 263
duplication (replication), 236
dynamics, 44–45, 67, 167, 195, 208, 212
effective procedure, 23

Einstein, Albert, xvii, 25, 27, 34, 36–37, 81, 126n6, 193–195, 248
  completeness and, 7, 221–222
  deductionism and, 261
  EPR and, 193
embodiment, 48–49, 109–110, 112, 119, 124–125
empiricism, 277, 294
energy (concept of), 28–29, 90, 175, 213
  entropy and, 27, 29
  mass and, 29
entanglement, 32
entropy
  information and, 71, 75, 92, 95, 96n20, 132
environment, 232, 290
EPR (Einstein-Podolsky-Rosen), 193
equation of experience, 5–7
equilibrium (thermodynamic), 186, 207ff, 212, 216n27, 267, 280
ethics, 187
evolutionary epistemology, 34, 81
experience (phenomenal), 4, 8, 11, 105, 109, 115, 247, 279
explanatory gap, xi, xxn2, 105
extension
  spatial, 10, 17n10, 52–53, 195
  vs. intension, 89, 116n4, 233
external world, xi, xii, 4, 6, 8, 9, 44

fall (biblical), 45, 125
fashion (intellectual), 49, 55, 146
fatalism, xv, 155, 156, 166, 175
feminine principle, 298
fiat, 15, 30, 63, 83, 155, 188, 203
field (concept), 29
first-order science, 44, 49, 55, 104, 106, 146, 282, 283, 284
fitness (evolutionary), 114, 133, 136
  landscape, 143, 264
fluctuation (random), 14
force (concept of), 24, 27, 30, 46, 52, 231, 240–241
form (Aristotle), 20, 169
formalization, 79, 86, 141, 189ff, 218, 229ff
  formal (axiomatic) system, 15, 113, 140, 183–184, 190, 210

free will (vs. determinism), 59, 244
fundamentalism
  religious, 177

Galileo, 24, 25, 45, 46, 85, 160, 165, 174, 192, 195, 204, 215n10
game (as metaphor), 94
gender issue, 124, 178, 247, 300
genetic condtioning, xiv
globalism, 149
goddess religions, 170
God
  hypothesis, 59, 137n4, 194
  rational, 47, 176, 184, 187
Gödel, Kurt, xvii, 32, 74, 85
governing power, xv, 47, 50, 63–65, 120, 222, 252, 268, 281
gravitation
  anomalies, 27
Greek heritage, xiv, 21, 155–156, 170, 177

halting problem, 237n4
hard problem (of consciousness), xi, xxn2, 102, 111–112
Heraclitus, 87, 208, 210, 213
heroism (masculine), 123, 134
holism, 277–278
homeostasis, 255–256, 269, 281
human-centric, 256, 282
humanism (*also* humanist), 124, 175
Hume, David, 28, 63, 65, 167, 194, 196n3, 231, 293, 297

idealism, xii, 7, 54ff, 123, 126n3, 201, 223
  descriptive, 201
  natural, 202–203
  normative, 201
  philosophical, xi
idealization, 201–202
identity
  of indiscernibles, 33, 201
  individual, 90, 212–215, 232, 278
  of particles, 212
I/it (relationship)
  vs. *I/thou* (relationship), 288
image (mental), 218, 225n3
imagination (role of), 14, 85, 175
immanent reality, x, xxn1, 64, 167, 169–170, 217–218, 222, 244–245, 255, 264, 285, 294, 301
impenetrability, 35–36, 213–224
Indo-European (languages), 15

307

Industrial Revolution, x, 150, 254
inertness (of matter), 43, 109, 240, 252, 253, 276
infinity (orders of), 38
information
  biological, 249ff
  black holes and, 94ff
  bound, 91, 94–95
  meaning and, 90ff, 112, 161–162, 223
  ontology of, 27, 93ff, 223
  processing, 27, 93, 109, 147, 249
  technical definition of, 91
initial condition, xv, 23, 60, 73–75, 131, 170, 195, 200, 206–207, 211–212, 225n6, 247, 259–260, 262–267, 272n29, 281
Inquisition, 46
intelligence
  alien, 63, 93, 204, 224
  artificial, 114, 125, 234–235, 256
  evolution of, 81, 287
intelligent design (movement), 250
intension (vs. extension), 17n10, 90, 202, 215, 233
  and intention, 17n10, 116n4
intentional(ity), 10–11, 21, 62–63, 111–115, 184, 187, 245–247, 251, 253, 290
  connection, 111–112, 115
  stance, 257n5, 304n4
  system, 109ff
interpretation (of formal system), 113, 190
invariance (and symmetry), 201
inverted spectrum (argument), 116n6
isolated system, x, xiii, 32, 75, 88, 186, 192, 204, 206, 212, 241, 243, 251, 256, 270, 280, 290

Kant, Immanuel, 8, 34, 41, 229, 262, 268
Kepler, Johannes, 26, 45, 192, 240, 263
kind (natural), 134, 162, 184, 248

law (jurisprudence), 46ff, 156
  divine, 60–61, 69, 71, 169, 171, 220, 247
  empirical, 64–65, 67, 167, 183, 191–192
  meta-law, 262, 266, 268
  natural, 25, 64–65, 167, 183–184, 188, 191–192, 195, 200, 222, 261–262
Leibniz, Gottfried, 26, 33, 74, 79, 105, 114, 254
linear(ity), 67, 233, 277
localism (vs. globalism), 121
logic, 80, 96n4, 113, 134, 156, 174, 242
Logos, 155–156

machine
  concept of, 15, 21, 23, 81, 202
  as metaphor, 118, 162
  nature as, 118, 159, 187
  organism and, 110, 239–240, 244, 250, 255
  as formalism, 162
  as metaphor, 251
  nature as, 253, 254
  organism and, 244, 250
  universal (computer), 23, 26, 233
macroscopic realm, 7, 16, 73, 89, 134–135, 213
maker's knowledge, 185, 186, 187
map (vs. territory), xv, 16, 68, 132
masculine
  principle, 123, 296
  world, 123
mass (concept of), 29, 95, 170, 219
materialism (philosophical), 43
mathematical universe hypothesis, 147, 223
mathematics
  domination by, 24
  effectiveness of, xiii, 79–82, 251
  nature and, xvii, 81–83, 191, 192, 195
mathematization, 69–73, 83–84, 201
Maxwell's demon, 131, 132
measurement problem, 17n8, 17n13, 54, 95, 134, 208, 284, 286
mechanism (philosophy of), 21, 23, 118, 147, 167, 200, 239–240, 242, 244, 250, 253
mechanist metaphor, xiii, 133, 202
medicine, 157, 158
medieval (thought, mentality), 160, 192
metaphor (role of), 3, 133
metaphysics, xiv, 54, 158, 283, 295, 303
microscopic realm, 7, 12, 82, 89, 134, 214, 225n10, 239
mind-body problem (MBP)
  as larger issue, 101–102
  solution of, 102
model (conceptual)
  mathematical, xiii, 27, 43, 50, 82–83, 136
modernity, 44, 178, 303
monoculture, 146, 301
Monopoly, (game of), 190
monotheism, 170–171
moral concern, 288
mortality, 119
Mother Nature, 291

308

Index

myth(ology), x, xv, 41, 48, 124, 139, 141, 273

nanotechnology, 235
Nasrudin, xv, 144
natural rights, 288–291
natural selection
   cosmic, 260, 264–267
natural standpoint, 8–9
n-dimensional space, 86ff, 96n11, 264
necessity, 60, 69, 183, 231, 253
   causal, 63, 65–66, 231
   logical, 61, 156, 184, 189, 190, 220, 260
   metaphysical, 65
   natural, 184
neo-platonism, 170, 204
neural processing, 4–5, 106, 110, 133
Newton, Isaac, 24–25, 28, 37, 46, 50, 61, 123, 126n8, 130, 160, 166, 170–171, 194–195, 240–241, 248, 260
   Principia, 219, 261
noise (statistical error), 14, 33, 51, 54, 75, 199, 296
nomological machine, 38n6, 204–206
nonlocality, 32, 36, 274, 278

object
   conceptual, 48
   perceptual, 12
   scientific, 19
objectification (see also *reification*), xviii, 8, 47–48, 84, 90, 103, 287
objectivity
   concept of, 6, 52, 55, 104, 284–285
   evolutionary advantage of, 104
   ideal of, 42, 162, 176, 201, 283, 295
   pretense of, xv, 55, 104, 203, 285
   subjectivity and, xiii, 10, 49, 91, 103, 105, 187, 263, 282–283, 285
observer (epistemic), xiii, 5–6, 9, 22, 36, 52, 55, 87–89, 96n14, 115, 143, 173, 200, 220
   external, xvii, 5, 106–107, 110–112, 135, 281–284, 295
   idealized, 130ff
Occam's razor, 200
omniscience, 7
ontology, 27ff, 102, 202, 254, 276, 301
   of information, 90, 93–95, 147, 249
   of science, 31, 34, 36, 47, 54, 66, 144, 195, 265–266, 284
operationalism, 9

order (vs. chaos), 21, 34
organism
   artificial, 224
   vs. mechanism, 21, 235–236, 239–258, 270
   metaphor of, x, 239, 256

pain (meaning of), 10–11, 101, 114–115
pancomputationalism, 39n16
parameter (theoretical), xix, 23, 191, 193–194, 199, 262–267, 270–271
passivity (of matter), 200
patriarchy, 170–171, 298
pattern, 12–14, 24
Pauli exclusion principle, 36
person(concept of), 102ff, 288
   first-, 103–105, 110, 113–114, 245
   law and, xix, 289–290
   nature as, 276, 288–290, 292n28
   second-, 104, 107
   third-, xvi, 43, 103–105, 109, 111–114, 245, 276, 278
phenomenal world, 4, 10–11, 109, 115, 210, 247, 279
phenomenology, 8, 10
   and Husserl, Edmund, 7–10
philosophy
   of mind, xix
   of science, xix, 43, 106
Planck unit, 93
Plato, xii, 15, 20, 169, 177, 201, 218–219, 251, 294
   platonism, xvii, 61, 83, 85–86, 167, 171, 189, 192, 195, 203–204, 223, 262
plenum (vs. void), 35, 218
point of view
   epistemic, 4, 59, 106, 109–114, 119, 135, 244, 256, 276, 283
   in language, 105
posthuman(ism), 287, 291
postmodern(ism), 150
power (quest for), 296
preestablished harmony, 33–34, 79, 81, 114, 204
principle of sufficient reason, 33, 201
printing (social role of), 124, 157–158, 165, 172
prior probability, 70, 72, 76n14
problem of cognitive domains, 132–134, 142, 286
product of definition, 24, 53, 70, 85, 93, 141, 159, 161, 189, 230, 252, 254

309

progress (ideology of), 175
projection
   perceptual, 8–9, 48, 202
   psychological, xvii, 21, 33, 62–63, 65, 80, 133, 201–202, 241, 245–246, 250–251, 288, 297, 304n6
prophecy, 161
propositional knowledge, 230–233, 65, 183–184
proton, 214, 256
psychologist's fallacy, 110–111
Puritanism, 177
Pythogorean(ism), 177, 192, 219, 223, 262
   theorem, 87

qualia, 115
quality (phenomenal), 10–11, 109, 115
   primary vs. secondary, 10–12
quantum (physics)
   microscopic realm of, 7, 12
   weirdness of, 6, 35, 92, 222

randomness, 27, 62–64, 76n3, 194, 218
   mathematical treatment of, 69ff, 73ff, 92
   pseudo-, 68, 74, 222, 230, 237n5
   unprovability of, 222
rationality
   and game theory, 151n17
   and God, 176, 185
   of nature, 194
realism, 220ff
   and idealism, 49, 54–56, 202
   naïve, 26, 202, 287
   natural, 6, 121
   scientific, 9, 19, 28, 79, 94, 135, 193
reality
   derived vs. original, 254
   immanent, 64, 217–218
   and mind-independence, xi, 6, 14, 20, 25, 203, 220, 255
realizing faculty, 8
realness (as phenomenal quality), xii, 12
reductionism, xii, 11, 25, 50–53, 106, 143
   deductionism and, 21, 190, 262, 277
   holism and, 49, 232, 255, 277–278
reference frame, 209, 210, 213
reflexivity
   consciousness as, 105, 285
Reformation (European), xiv
reification, 15, 27–32, 94, 201, 204, 274

relativity
   general (GR), 37
   length contraction vs. time dilation, 36
   special (SR), 36
representation, xvii, 7, 105, 109ff, 234
   cognitive, 48, 112, 114, 156
   mathematical, 229
Restoration (English), 47, 159
reverse-engineering, 186, 235
reversibility (of time sig.), 186, 207ff, 266
   reversible system, 75, 161, 186, 191, 207, 210
revisionism (scientific), 42
Royal Society, 47, 140, 149, 159, 176

scholasticism (medieval), 157, 160
scripture, 155, 157–158, 160
second law of thermodynamics, 71, 131, 207, 265
second-order science, 281–286
secularism, 177–179
self-consciousness (reflexivity)
   evolutionary role of, xii, 104
self-definition, xvi, 77n26, 246
self-design, 187, 246, 247
self-existence, xxn1, 20
self-organization, xiv, 220, 240, 242, 243, 246–248
semantic meaning (vs. syntax), 162
sense data, 4
sentience, 102, 107
simplicity (ideal of), 55, 76, 191, 197n22, 200–201, 240, 250, 287
   simplification, xviii, 190, 205–206, 240, 252
simulation, 63, 93, 197n28, 202, 222ff, 234
   argument, 223–225
   vs. reality, 202, 217, 222ff, 232
singularity (black hole), 216n22
skepticism
   cognitive role of, 9, 279, 299
   philosophical, 8, 168, 178, 185, 217, 223, 294
space
   conceptual, 52, 86–88, 264
   geometrical, 86, 200, 209
   time and, 54, 87, 200, 209
stance of unknowing, 297–303
standard model (of particle physics), 7, 144
statistical basis of k., 9, 67–69, 195, 293

Index

statistics (classical vs quantum), 73
string theory, 23
structure
  definition and, 50
  information and, 75, 90–95, 218
  mathematical, xii, 24, 27, 189, 243
subjectivity, 48, 102, 105, 109, 130, 168, 282, 287, 293–294
  evol. role of, xii, 9, 104, 279, 285
  v. object., 10, 42, 73, 115, 187, 193, 284, 295, 283, 295
  rise of, 168, 294
suffering, 102
superstition, 69, 165, 219
surrender (psychological), 300
symbol (role of), 14–15, 112, 115, 121, 157, 185
symmetry (vs. asymm), 84, 192ff, 201
  breaking, 64, 84, 192ff, 253
syntax (vs. semantics), 15, 92, 185

technology
  salvation and, 125
  science and, 122, 175, 176, 297
  society and, xiv, xviii, 149, 175–176, 178
teleology, 22, 250–253
text
  as deductive system, 161
  exegesis of, 155, 159, 172
  nature as, 85, 155–156, 159–160, 172, 304n6
  scientific theory as, 162
textualism, 160–161
thema (G. Holton), 17n7, 49, 53, 186
theology, 177, 184, 220, 295
  theory and, 31, 43, 142, 167, 244, 298, 300
theory-dependence, 33, 291
theory of everything, ix, xvii, 259
thermodynamics, 212, 280
third-order science, 287
time
  asymmetry, 266
  beginning of, 37, 133, 268
  dilation, 36
  meta, 277
  reality of, 56n5, 191
  space and, 54, 87, 200, 209, 262
top-down, 125–126, 234, 245–246, 250
transcendence, 50, 119–120, 122, 124, 170, 247
transubstantiation, 168
truth (vs. proof), 183ff, 189, 193
Turing test, 235
Twenty Questions(game), 93–94

uncertainty, 69, 73, 117, 129, 293–294, 296–297, 300–301
  Heisenberg's p. of, 55, 208, 130
undecidability, 232
unification, 7, 30, 35, 37, 146, 215, 279
universal machine (computer), xiii, 23, 233
unknown, the, 43, 125, 294, 297–298, 300–302
uploading (of mind), 236

variable
  in equation of experience, 5
  laws of nature, 264, 267
  hidden, 89, 96n15, 194
  physical, 33, 83, 95, 133, 208, 211, 221
verum ipsum factum (Vico), 187
Vico, Giambattista, 185, 186–188, 248, 294
virtual reality, 93, 217, 224
visual sense, 11, 51–52, 144

vitalism, 244
void (vs. plenum), 35, 218

war
  and culture, 117, 121, 150
  on nature, 121ff
watchmaker argument, 251
wave-particle duality, 16, 35, 37, 274
wilderness (the wild), 20
world
  of a game, 48, 119, 126, 141, 190
  possible, 7, 143, 188, 266, 268
  reality of the, 6, 178
  variable (in equation of experience), 5
world-in-itself (Kant), 300
writing (vs. speech), 156ff